Talking About Relativity

Talking About Relativity

By

J.L. SYNGE

To Denis Johnston,
playwright and author,
who seemed to want
something like this.

1970

NORTH-HOLLAND PUBLISHING COMPANY
AMSTERDAM · LONDON

AMERICAN ELSEVIER PUBLISHING COMPANY, INC.
NEW YORK

Publishers:

NORTH-HOLLAND PUBLISHING CO.–AMSTERDAM

Sole distributors for the U.S.A. and Canada:

AMERICAN ELSEVIER PUBLISHING CO., INC.
52 VANDERBILT AVENUE, NEW YORK, N.Y. 10017

Library of Congress Catalog Card Number: 70 140485

ISBN North-Holland: 0 7204 0206 9
ISBN American Elsevier: 0 444 10076 8

PRINTED IN THE NETHERLANDS

Preface

From time to time young men have brought me, for my advice, pieces of work, pretty well completed but with no title. This amazes me, for to me the title is the compass-setting by which the whole development is steered. Of course, I may change it on second thoughts, but then whatever emerges in the end is schizophrenic – one cannot serve two masters.

The contents of the present book are distilled from cogitations extending back fifty years. Copious draughts of this distillate were made available to captive audiences in several universities and to the world at large in the more conventional books I have written about relativity. Some of the particulars were presented in lectures a few years ago at the University of Kentucky and University College, Dublin. To all who have listened to me or have read at least some of what I have written, I extend grateful and sincere thanks.

This book may fall into the hands of people who have never listened to lectures in a university or to papers presented at scientific meetings. I have done both, emerging in each instance a little dazed, usually filled with admiration at the erudition of the speaker, but regretful that I was able to understand so little of what he said. Innate modesty told me that the fault was mine and that I should not feel misused provided the words were audible and the formulae visible. But modesty can be pushed too far. A decisive event for me was when I heard Schrödinger say good-humouredly after a lecture by a colleague: 'Do you know, I didn't understand a single word of what you said.' It gave me good excuse for saying to myself (but not to him) after a lecture by Schrödinger that I didn't understand a single word of it. Another decisive event was reading in a book by F.D. Ommanney* a generic description of scientific meetings which it would be hard to better.

I started this Preface by talking about titles and their importance. I shall be getting back to that in a moment, but first I must correct an impression of disappointment and frustration which it was the intention of the preceding paragraph to create. Communication in scientific matters may be bad, but it exists. It must, or there would be no science to communicate. It does indeed exist, and of all forms of communication it is the most exciting when it succeeds, the most depressing when it fails. But it does not take place through the formal channels prescribed for it, at least not nearly to the extent that many people suppose. If among the semi-inarticulate demands of the present generation of revolting students there is contained a demand to clean up the channels of communication, then I am for them in principle.

But it is not easy, this task of communication. I start off

* F.D. Ommanney, *The River Bank* (Longmans, London, 1966), in particular pp. 216–220.

on the wrong foot by speaking of a task. It creates the wrong atmosphere. The human mind is a wild and wayward thing, resenting discipline unless self-imposed. I take the view that, generally speaking, people want to understand things they do not at the moment understand, and that other people, who think they understand these things, are only too anxious to spread their understanding to those who seek it. The problem is, how to do it. One does not of course decry the elaborate educational systems existent in all civilised countries. They exist for this very purpose, but with overtones (economic, political, national and so on) which tend to mask the simple human fact that what is sought is not really economic prosperity or prestige (national or personal) but understanding, pure and simple – the religion of modern man.

For me at least, the monologue, spoken or printed, is of small value as a vehicle for communication of ideas. Dialogue is essential, and the simplest form of dialogue is conversation between two people, or perhaps three or four, certainly not a crowd. But at a pinch one can talk to oneself or one can try to exchange thoughts with the author of a book with the page open before one. These are the ways in which thought really grows, and not through the formal channels. I chose the title of this book before I wrote a word of it as a protest against a dark shadow which has hung over academic life for centuries – the shadow of authority, of pedantry, of all the foolish high-sounding non-sense (to the auditors) applauded because applause is called for – 'polite applause,' as Ommanney writes, 'like a brief shower of rain on a roof.'

So I called this book 'Talking about Relativity', each word pregnant, at least to me. 'Relativity' – obvious, this execrable word for which we can find no alternative. 'Talking' – yes, I have tried to imagine in writing that the floor is not mine, but that you are there, ready at a moment's notice to protest against some folly of mine or to fall asleep through sheer boredom. As for 'about' – well, it is about it and about, no fixed goal, no examination to be passed at the end. Now that the book is written and I look it over, I realise where it lives up to my intention and where it falls off into dogmatism and pedantry. But I am in no position to judge. To a writer, a finished book is about as interesting as a dead fish is to the fisherman who has played it. Only time will tell. If the writer comes upon his book after he has forgotten what is in it, he may find that it stinks to high heaven, or he may say with Swift 'What a genius I had when I wrote that!' Only time can tell.

This is a book about concepts, and even now that it is finished I don't know what a concept is. But I have invented some terms which may be useful. I have made a sharp distinction between the real world (*the R-world*) and the several model worlds (*M-worlds*) created by the mind of man. If I could, I would give credit to those who invented this distinction. Also, I have invented the name *Pygmalion syndrome* for that disease of the mind which blurs the distinction between R-world and M-worlds. I may be accused of burdening physics with unnecessary terms, but we must have words in order to communicate with one another, and in the present connection some such words appear to me absolutely essential. Since I named the Pygmalion syndrome I have been putting every theoretical physicist I meet up on the X-ray slab and examining his innards: the disease is rife.

Any library that buys this book will be faced with the

problem of classifying it. I would propose PHYSICS. True, there is not very much physics in the book, but what there is is very, very fundamental. Although not by any means a textbook, it offers, I hope, a sound foundation – much still to learn but nothing to unlearn. If the library can rise to a second copy, it might go under PHILOSOPHY, a discipline now sadly divorced from physics, with which at one time it was almost coextensive. If there were a question of a third copy, PHILOSOPHY OF SCIENCE would be the place. But although I treasure the fact that Schrödinger in his last book* (published posthumously) described me as 'a very amusing conversationalist', I do not think that, in spite of the title, that should be made a basis for classification.

But to quit fooling, I would like you to realise that I have faced a problem of enormous difficulty, a problem of communication. A university lecturer gets to know his class and, if wise and good, sets the tone so that it is rather too easy for the best to follow and rather too hard for the worst. A lecturer who goes to a strange place to give an isolated lecture or two is in a much worse position, for any inquiry about the absorption-coefficient of his audience is likely to get an answer which, through local vanity, exaggerates it. But the author of a book is in the worst possible position for he can make only the wildest of guesses about the reception it is likely to receive. What is clear to one reader is obscure to another. What is dramatic to one is banal to another. What is amusing to one is stupid to another. And so on. Therefore I have incorporated a variety of styles in the hope that you, like the proverbial curate, will find the egg good in places – to ask more would be to ask too much.

* E. Schrödinger, *My View of the World* (Cambridge University Press, 1964).

Contents

CHAPTER I

Talking about concepts

The aim of this book is to present the basic concepts of the theory of relativity in a form comprehensible to a wide public of educated people who would like to know more about a subject wrapped in mystery except for a comparatively small group of professional relativists. It may be that these last will not wholly approve of my ideas, for concepts are tricky things, but I believe that relativity, as I present it here, is not seriously out of line with what is commonly accepted by experts.

However the book is not written for them. I aim at those who have never received any organised instruction in relativity and also at those physicists and other scientists who have had some such instruction and have occasion to use some relativistic formulae, but in a puzzled way, not knowing what lies behind them.

As a problem in communication, my task is fantastically difficult, and, while I don't wish to sound a note of despondency, I don't want to sound unduly optimistic either. There is first the question of mathematics. What level of mathematical competence am I to assume in the reader? If I assume too high a level, I shall leave him bored and frustrated, and the main purpose of the book would not be achieved. On the other hand, if I attempt to get along without mathematics, relying on vague talk, only an unsatisfying vagueness can result. So I am going to compromise, making a wild guess at the point where the reader is likely to get lost and running over some aspects of mathematics likely to throw light on the main theme.

In giving a mathematical lecture it is customary to write formulae on a blackboard, to the accompaniment of a verbal commentary which includes the translation of the symbols into words. The lecturer writes x^2 and says 'ex squared'. He might dispense with the blackboard and rely on the spoken word were it not for the limitations of the human mind. We find it easier to concentrate on a thing when it is displayed to us in some concrete form. Nevertheless it may be asserted that mathematical symbolism is merely a very efficient shorthand for recording what might be said or written out at length in words.

To do mathematics, we have to learn that shorthand. But that is by no means the whole of mathematics, for, whether we use symbols or spoken words, we are dealing with mathematical concepts all the time except when we are launched on a piece of routine calculation which we can carry out mechanically. You ask me what a concept is. This stumps me, for I can only offer a definition involving other words and these in turn will demand definition, and so we shall never stop. The best plan is to read on, and perhaps you will begin to see what I mean by the word *concept*.

To get the ball rolling, let us start on an everyday level. It is a question of communication, really. If I say to you

‘Please sit down in that chair’, you may or may not do as requested, but at any rate you know what I mean by the word *chair*. You have, in fact, the concept of a chair in your mind. When you were born you didn’t have that concept. It grew up in you as the result of experience, and it is now available for purposes of communication with other people, provided they share the concept.

It is not essential that a concept should be perfectly clear. It is not essential that you should be able to distinguish a chair from a stool that closely resembles a chair. As a matter of fact, there is always some fuzziness about a concept, and one of the main differences between ordinary life and science is that scientific concepts are less fuzzy. But they are fuzzy nevertheless. It is only in mathematics that we find clear-cut concepts, and it is probably this inhuman characteristic that makes the subject repellent to many people.

You might argue that understanding concepts is merely a matter of understanding what words mean. But that, I think, would be putting the cart before the horse, for the concept lies deeper than the word (or words) you use to refer to it. A child has the concept of a chair before he learns the word (which might be in any language, but with the concept essentially the same). Philosophers argue about the meaning of meaning, but I want to keep out of that deep water by thinking in terms of common experience, and I suggest that you look over a page of a dictionary, noting those words which you understand and those you do not understand – equivalently, those words which are matched by concepts in your mind and those words not so matched. When you examine the second class, you will find them broken into two subclasses: first, those words for which the dictionary supplies an explanation evoking a concept you already have; secondly, those words for which the dictionary explanation leaves you as much in the dark as before. You do not, as a rule, enlarge your supply of concepts by consulting a dictionary – you merely get more words to talk about those concepts you already have.

But from time to time new concepts are created in your mind. The greater part of this creation took place in early childhood, but the process goes on through life, and this book is supposed to be making its contribution. I have chosen its title because I believe that concepts are not things you can rush at and grab. They come to you by indirection, by talking about things, much more easily than by direct attack. The mind has a way of freezing up when faced by orders to understand this or that, and it is necessary to get into a suitably relaxed state. I realise that advice to relax may be self-defeating – I never succeed in defeating insomnia by telling myself to relax – and perhaps I should not give such advice. Nevertheless I think the lines of Clough are apposite in regard to the acquisition of concepts:

> Ahead the sun climbs slow, how slowly,
> But westward, look, the land is bright.

As indicated earlier, I hold that concepts lie much deeper than words, and indeed men 'think with their hands', gaining concepts without necessarily the words to describe them. But I must rely on words, such as they are, and here I have to be careful. To introduce concepts new to you, I must use only words which already evoke concepts in your mind. But in doing this, I must choose my words with quite extraordinary care, for of all languages English has the greatest supply of words and by choosing unsuitable words I may fail to hold your interest. Some words make dull reading. Others rouse a meretricious excitement and deflect attention from the proper business. I shall try to steer a middle course.

Talking about words, what about the words 'theory of relativity'? I believe that they possess a strong emotive attraction. Take the word *theory*. It is a dignified word, but it carries the implication that this is only a theory, a speculation, not a fact, and keeps us guessing whether it is true or not. As for *relativity*, it has a polysyllabic mystery in it, and makes one wonder what is relative to what. I often find myself wishing Einstein had chosen some other title, for the theory is fact (in a sense I shall go into later) and you have sometimes to search hard to find what is relative to what. But the title is now so widely accepted that it would be pedantic to change it for another and so I fall into line, suppressing my grumble.

Getting back to concepts, they may be hard to acquire but they seem much harder to get rid of. You might think that it was a case of the more concepts the better, but it is not always so. Two concepts may fight like hell with each other and reduce their owner to a frantic state. That happens when you introduce relativistic concepts into your mind without first purging it of Newtonian concepts, and this purging is almost impossible to achieve entirely since we have all grown up with Newtonian concepts of space and time, and most of the scientific world still uses those concepts.

The usual pattern of education in these matters is as follows. One starts by learning Euclidean geometry and Newtonian mechanics, and these drive home the Newtonian concepts which (although Newton made them more explicit) have been in the minds of mankind for a very long time. From that one may go on to study the special theory of relativity, regarding its concepts as modifications of Newtonian concepts. The final step is to study the general theory of relativity as a modification of the special theory. That plan follows the historical development and there is much to be said for it. I expect that the education of all relativists was along those lines. Mine was. But I suffered from considerable mental confusion about concepts. As the change was gradual, there seemed to be no decisive moment at which to throw out old concepts and replace them by new ones, and it took me many years to get a clear view of the concepts of relativity.

In the days when relativity had to win credence in an incredulous world, it was natural to give it respectability by explaining it as far as possible in terms of the old concepts. But this led to getting the concepts mixed up. Those days are past and one can take a new look at the problem of introducing relativistic concepts. It seemed to me that one should seek, not a gradual weaning, but a striking break, and so I resolved to take the reader right out of the domain of Newton into the most general aspects of relativity. In other words, to put general relativity first and special relativity last.

But I want to be friends with everyone. Even if I think that Newton was wrong and Einstein right, I want to interpret the words *wrong* and *right* in a certain way. Therefore I have coined the term *R-world* to refer to the *real* world, the world of immense complexity in which we live and move and have our being. In contrast we have various *M-worlds*, with M standing for *model* or *mathematical*; it matters little which word we use since it is in practice impossible to describe a model world without using some mathematics.

Suppose we use M_1-world to denote the Newtonian model and M_2-world to denote Einstein's model. Neither of them is *right* in the sense that it is an exact image of the R-world. No M-world can be that. But both of them are good images of certain features of the R-world, leaving many phenomena that they are both unable to deal with. Of the two, Newton's M_1-world is the more versatile; it can deal with a great variety of problems in technology, physics and astronomy, and deal with them with sufficient accuracy. Einstein's M_2-world involves us in harder mathematics and consequently is more difficult to apply. But where the two M-worlds overlap – where they both claim to predict occurrences in the R-world – it seems to be generally agreed that the M_2-world is better than the M_1-world. However the issue is often difficult to decide because, although their basic concepts are very different, their predictions often agree so closely that it is hard to decide which is the more correct.

I offer the concepts of R-world and M-worlds to get rid of a lot of mental confusion. But mental confusion is not always to be deplored. It is more exciting for a physicist to think that he is at grips with the realities of the R-world instead of indulging in the intellectual exercise of building an M-world. Indeed when he gives advice about the construction of apparatus, he is advising about the behaviour of the R-world. (In an M-world the apparatus is of a peculiarly idealised sort.) Thus physicists are very much inclined to think of their M-worlds as being identical with the R-world, and on the whole this is not a bad thing. But it is bad to do this without acknowledgement. It may lead to much confused argument, and it is well to have a name for this sort of confusion. I name it the *Pygmalion syndrome*, after Pygmalion who carved a statue of such surpassing realism that it came to life – M-world becoming R-world in the mind of the enthusiastic physicist.

At this point I have to ask myself whether I am carrying you along with me. It is so easy to lose contact. I had better do some more talking about the concepts of R-world and M-worlds and hope for the best.

If I dream about a man in a tall white hat, I may tell my wife in the morning about my dream, but neither of us believes that such a man was in our bedroom. He was not part of the R-world, but the dream was, and I regard the dream as being a matter of the electrochemistry of my brain, which brain, and what goes on in it, being part of the R-world. Men in tall white hats may, and probably do, exist in the R-world. To meet one in the street would cause only a mild surprise because the costume is unusual.

In the R-world, as I understand it, there is nothing supernatural. It would be a contradiction in terms since the R-world embraces all nature.

All experiments are carried out in the R-world. They are also designed in the R-world in the sense that the brain and hand of the designer belong to the R-world.

The purpose of physics through the ages has been to explore the R-world, and in that exploration there are two essential tools – the hand and the brain. The brain tells the hand what to do, and in return there is a feed-back from the hand to the brain. All this is, strictly speaking, in the R-world, but it is convenient to pick out some of the activities of the brain and say that they form an M-world. Such an M-world is, as Wordsworth wrote,

> In verity, an independent world,
> Created out of pure intelligence.

Thus, although every M-world is, in the last analysis, part of the R-world (a matter of the electrochemistry of the brain), it is convenient to regard it as something different – something that we can compare with the R-world as revealed by experiment.

Although I have coined the terms R-world and M-world for this book, the idea is far from original. I don't know how far back it goes, but it has become more familiar during the present century. Nevertheless I believe that many people suffer from the Pygmalion syndrome and that I have a duty to shake them out of it.

Happily I am talking to people with some acquaintance with the M_1-world of Newton and all I need to do is to remind you of the wheels which are perfect circles and the inclined planes which are perfectly smooth, and also to remind you that you use mathematics to solve problems in mechanics. All this is M_1-stuff. It would of course never have been created if the predictions obtained by reasoning in the M_1-world had proved completely unreliable when compared with experiment, but this verification of theory by experiment is a trivial matter in the present connection. The essential point is that if you studied Newtonian mechanics, even in a quite elementary form, you played in the M_1-world, and consequently it would be quite ridiculous for you to say that you didn't know what an M-world is like. Of course you do.

But it is wrong to try to bully a person into understanding. One must let things sink in. Give your mind a rest and let the subconscious take over. Perhaps, just before you go to sleep, a concept will emerge from the subconscious into the conscious, and if in the morning you find it still there, and not looking nonsensical, you may be able to call it your own. A concept will have been conceived.

There was a young woman called Wong
Who thought contraception was wrong.
Said she 'It is right
To conceive in the night
And to welcome the dawn with a song!'

CHAPTER 2

Talking about geometry

Geometry affords the simplest approach to the question of R-world versus M-world, and I would be suprised to learn that you have never suffered confusion in this regard. I have not sufficient insight into the minds of the ancient Greek geometers to known how they thought about this matter, but Euclid succeeded in confusing the issue for some twenty centuries – if he did not suffer from the Pygmalion syndrome himself, he did little to save others from it. It was only towards the end of the nineteenth century that Hilbert (with others) cleaned up the situation by splitting geometry into an M-geometry and an R-geometry, the first based on another M-world, that of numbers (of which more later), and the second left to draughtsmen, surveyors, astronomers and so on, although these last have been so completely victims of the Pygmalion syndrome that, even to the present day, they hardly realise that a revolution in human thought has taken place.

To explain all this, let us go back to the beginning of education in geometry. You are provided with paper, pencil, ruler and compasses. You draw figures and measure lengths and angles, and all this is definitely in the R-world, as real as your paper, your instruments, yourself. If you draw the bisectors of the angles of a triangle and find that, as far as you can see, they intersect at a point, this is an experimental result. It would be rash to draw conclusions from a single experiment (perhaps there was something peculiar about the triangle you used), but, if you find the bisectors intersecting for twenty triangles, you probably conclude that it is true for all triangles. You might also try the experiment using a much sharper pencil, or you might substitute light rays for the straight edge of the ruler, without shaking your conviction that the bisectors meet in a point, always.

That conviction is not unreasonable under the circumstances, but actually no reasoning was involved. You simply reported what you saw in the R-world. Reason enters only when you construct an M-world out of points, lines and circles, these concepts being derived by a sort of limiting process from concepts you already have of certain marks on the paper. An R-point is a smudge of pencil on paper; the corresponding M-point is derived from the R-point by forgetting its size and shape and concentrating on its position, so that an M-point has position and no other attributes. Likewise the M-line and the M-circle are derived from the R-line and R-circle by depriving them of their thickness. Unfortunately it is not usual to use different names for different things, and if you meet the words *point*, *line* and *circle*, you can only guess from the context whether they are R-things or M-things – in fact geometrical language seems designed to give you the Pygmalion syndrome. Punning is good fun, but in such cases it can lead to headaches.

The M-world of geometry has more in it than the concepts of point, line and circle and others like those. It contains also certain rules of procedure, called axioms or postulates. I must insist that these are rules of procedure rather than self-evident truths; they are more like 'Keep to the left' and 'Do not spit on the sidewalk' than 'A is either the same as B or different from B'. Whatever you call them, and however you think about them, the rules of procedure are essential. Without them, M-geometry cannot get off the ground, for its concepts cannot be handled or pushed about by human hands. You can cut an R-triangle out of paper with a pair of scissors and paste it up on the wall, but to think of any such action with an M-triangle would be absurd unless you first say what you mean by M-paper, M-scissors, M-paste and M-wall.

Prior to Hilbert, the Pygmalion syndrome was rife among geometers. They confused the two worlds and thought that they were proving things in the R-world by proving them in the M-world. This was understandable since things proved in the M-world did in fact turn out to be true in the R-world, at least to the order of accuracy available. The bisectors of the angles of a triangle were proved to meet in a point and did so, the perpendiculars dropped from the vertices on the sides were proved to meet in a point and did so, the sum of the angles of a triangle was proved to be two right angles and it was. Except for a few isolated doubters, everyone was happy.

The confusion between R-geometry and M-geometry was not brought to an end by any crucial experiment in which an M-deduction contradicted an R-experiment. It was dissipated by a purely intellectual effort, an extremely difficult effort for anyone suffering from the Pygmalion syndrome which inhibited any attempt to separate the two worlds. Since I am using geometry, not for its own sake, but as perhaps the best illustration of R-world versus M-world, I shall not attempt any detailed account of the history of this essential separation. But I would like to refer to the angle-sum of a triangle, which I said was proved to be two right angles and in fact was found to be two right angles. As for the proof, it will be found in any textbook on Euclidean geometry, but it is easy to see that one of the rules of procedure amounts to the acceptance of that result from the start. The Pygmalion syndrome was somewhat shaken by the discovery of other M-geometries (non-Euclidean geometries) in which the angle-sum is greater than or less than two right angles. As logical structures, these M-geometries were just as good as Euclid's M-geometry and the question naturally arose whether R-geometry (the geometry of the real world) might not be better represented by one of those other M-geometries.

That was a salutary shock but it took many more years to shake off the Pygmalion syndrome in geometry. Indeed I doubt whether it has been shaken off yet. Mathematicians are certainly free from it, but I am not so sure about physicists, and as for the wider public I have the gravest suspicions. It is indeed quite strange that R-geometry and M-geometry of any type should be confused, for they are animals of quite different types, R-geometry being a matter of pencils, paper and geometrical instruments, with perhaps light rays brought in, whereas M-geometries are in the mind.

Let us now take a closer look at R-geometry, a game played with pencil, paper, etc. Imagine you are drawing geometrical pictures on a sheet of paper laid on the table in front of you. The table stands on the floor and the floor is carried by the earth which rotates about its axis once a day and goes round the sun once a year. Your R-geometry is the body of results recorded in a notebook, these results being experimental and not derived from any M-reasoning.

But might not another person under different conditions record different results? You might try the effect of circumstances by tilting your table. Your might move your table to a pole or to the equator. You might put it on a train or an aircraft. You might carry it in a spacecraft to the moon. You might mount it on a turntable and use a powerful motor to make the turntable spin rapidly.

Unless the records of experiments conducted under these various circumstances are in reasonable agreement with each other, then there is no such thing as R-geometry but rather various R-geometries depending on circumstances.

Now I do not want to suggest that R-geometry is highly sensitive to circumstances. If so, this would have been noticed long ago. But I would suggest that some sensitivity to circumstances is to be expected. Take the turntable, for instance, and suppose it is made to spin very fast, with some geometrical figure inscribed on its surface. If the turntable turns fast enough, it will be torn in pieces by centrifugal force, and there will be no geometry left to talk about. At a smaller speed, but still large, the centrifugal force will distort the diagram, changing the distances between points on it.

The answer to that is to inscribe the diagram, not on steel which yields under great stress, but on a piece of rigid material. But where is a rigid body to be found? All bodies in the R-world yield to stress, some more than others, but all yield. The mere mention of a rigid body is a sign that the Pygmalion syndrome is at work, for rigidity is a concept present in the M-world of Euclid and Newton but not in the R-world.

But concepts in the R-world are always fuzzy. The R-point is a fuzzy concept, but that did not prevent Euclid from using M-points in his M-geometry. In the R-world, rigidity is a fuzzy concept (a stone is rigid in comparison with rubber, for example), but this fuzziness did not prevent Euclid and Newton from putting the concept of rigidity into their M-worlds. To enlarge on this, let me say that you are not likely to find the word *rigid* written explicitly in a textbook on geometry, but if you think of transferring a distance from one part of a diagram to another part by means of a pair of compasses, you are in effect assuming that the compasses form a rigid body, not of course when you adjust the legs, but when you don't, deliberately. As for Newtonian mechanics, every student knows that, after he has studied the dynamics of particles, the next thing is the dynamics of rigid bodies, bodies which suffer no deformation whatever. From the standpoint of the engineer and the astronomer, it would be fatal to abandon the concept of a rigid body; without that concept, the flywheel and the spinning planet would present very messy problems indeed.

I have to assert very solemnly and seriously that the concept of rigidity is a dangerous concept. By that I mean that if you admit that concept unthinkingly into the M_2-world of Einstein, it will fight with other concepts – it will involve us in logical contradictions and nothing but confusion can result.

As I shall tell later in some detail, the M_1-world of Newton and the M_2-world of Einstein have some basic concepts in common. But the ways part, as of course they must, or the two M-worlds would be the same and there would be no relativity to talk about. There are two important forks on the road, but they lie so close together that they are hardly distinguishable. One fork is at the concept of *simultaneity* or absolute time (of which more later) and the other at the concept of *rigidity*. If you want to travel Newton's road, you accept those two M-concepts; if you want to travel Einstein's road, you don't accept them. Further along Einstein's road, there may be other concepts to accept, but that is another matter.

If you reject the concept of rigidity, you automatically reject Euclidean geometry. It may come creeping in again by the backdoor, but it is better to be a bit dogmatic at this point.

I feel that some sort of apology for this chapter is needed. I brought in geometry and now I throw it out. But I could think of no better way of introducing and explaining the contrast between R-world and M-world, and if I have succeeded in making those concepts clearer, perhaps no apology is needed after all.

CHAPTER 3

Talking about algebra

I am going to use the word *algebra* in a general sense to cover not only algebra proper but all mathematics in which letters of the alphabet represent numbers or operations on numbers. I want to compare algebra, thus understood, with geometry, as understood in the preceding chapter. I pointed out there that great confusion reigned until it was realised that R-geometry and M-geometry must be thought of as different things, and I went on to warn you that the M-geometry of Euclid is not to be carried forward into the relativistic M-world, except with great circumspection. Let us now try to see how algebra stands in regard to this business of R-world and M-world.

Algebra originated in the primitive operation of counting, an operation in the R-world. We record the results of counting in a notation familiar even to small children: 3 cats, 4 aircraft carriers, 7 beads, and so on.

We take two boxes containing beads, count the beads in each box and then make an overall count. The result is recorded as, say,

$$2 \text{ beads} + 3 \text{ beads} = 5 \text{ beads}\ ,$$

the notation implying that we counted the 2 beads before counting the 3 beads. But we soon realise that it doesn't matter in which order we count them: we have also

$$3 \text{ beads} + 2 \text{ beads} = 5 \text{ beads} .$$

In writing down these statements of experience, we have introduced two useful shorthand signs (+ and =), but they are no more than the equivalent of sets of words.

It soon dawns on us that it does not matter what objects we count: the result is essentially the same, and so, perhaps out of laziness, we omit the name of the thing and write

$$2+3=3+2=5 .$$

We are still in the R-world, or at any rate so close to it that we can run back into it immediately by putting in names after the numbers. A little care is needed, or we may find ourselves writing down some such nonsense as this:

$$\begin{aligned} &2 \text{ aircraft carriers} + 3 \text{ women} \\ &\qquad = 5 \text{ female aircraft carriers} = 5 \text{ flighty women.} \end{aligned}$$

It is clear that, in any such statement, we should use only one name throughout. Then we are safely back in the R-world.

This simple precaution fails, however, when it comes to multiplication. We know that $2 \times 3 = 6$, but what does it mean? To bring it into the R-world, we must put in names, and there are at least two ways of doing this. It may mean that, if you have 2 boxes, each containing 3 beads, then you have 6 beads altogether. Or it may mean 2 feet $\times$ 3 feet = 6 square feet, with reference to the area of a rectangle.

Our ancestors realised a long time ago that the situation called for the creation of some kind of M-world inhabited by numbers and certain signs, notably signs for addition, subtraction, multiplication, division and equality ($+$, $-$, $\times$, $\div$, $=$), this M-world being ruled by certain laws. But the process was long and confused, and it was only in the nineteenth century that this M-world made its Unilateral Declaration of Independence. In the minds of most people it is still a colony of the R-world, and this is a more excusable case of the Pygmalion syndrome than the identification of R-geometry with M-geometry. Indeed, in an oft-quoted statement, the mathematician Kronecker said that God created the positive integers and man created all the rest; translating this into untheological language, I would say that Kronecker put the positive integers 1, 2, 3,... into the R-world and negative numbers, fractions, irrational numbers, etc. into an M-world. I would go further and put even the positive integers into the M-world of numbers, admitting however that they are sitting on the fence.

As for this M-world, it would be out of place for me to attempt here anything in the nature of a systematic development, as has been done in very readable form by my colleague, Professor C. Lanczos, in a paperback entitled *Numbers without end* (Oliver and Boyd, Edinburgh 1968). But at the same time I must try to give at least a thumbnail sketch of the rock on which theoretical physics is built, for this M-world is indeed a rock far more reliable to build on than geometry, which, as I said earlier, must be treated with circumspection.

Although what I am going to talk about is, in the last analysis, of paramount importance to the human race (life and death are at stake), I suggest that the proper frame of mind is not one of intense seriousness but rather one with a large measure of playfulness in it. This is a game that all can play, irrespective of colour, creed, class, or economic condition, the apparatus required being merely pencil and paper, and even these can be dispensed with if your brain is in good order. You may imagine that you win a prize for every interesting true statement and pay a forfeit for every piece of nonsense. The basic rules of the game are the rules of ordinary logic, supplemented by certain other rules.

The game starts with the positive integers $1, 2, 3, \ldots$ laid out before us, and the signs $+$ and $=$.

I challenge you to find a number x such that $x+2=5$. You answer $x=3$ and win a small prize.

I next challenge you with $x+2=2$, and you give the correct answer that there is no solution. Although your answer is true, you get no prize because it is not interesting. To win a prize, you must make a big fuss and protest that, although there are infinitely many integers, you need one more, namely 0 (zero), and that the answer is $x=0$. You insist that the game must henceforth be played with $0, 1, 2, 3, \ldots$. Others realise that this is a good idea, and you receive a large prize.*

Much the same thing happens when I challenge you with $x+5=2$. There is no solution, but one must be made, viz. $x=-3$, and so we extend our number system

* It is an interesting fact that, since Nobel was jealous of the Swedish mathematician Mittag-Leffler (a man who seems to have thrown his weight about quite a bit), there is no Nobel prize in mathematics. Had Nobel prizes been in existence at the time, without this restriction, the inventor of zero would surely have received one.

to be

$$\ldots -3, -2, -1, 0, 1, 2, 3, \ldots .$$

Rules are set up for manipulating these numbers and the sign of addition (+). Since you know perfectly well what the result of any such operation would be, I shall not bore you with formal statements, but go on to the signs of subtraction and multiplication. But these are not of much interest in that they do not carry us out of the infinite set of numbers already before us. However you might pause for a moment to ask why we say that $(-2)\times(-3)=6$. It is futile to run back into the R-world and ask what stock of beads you would possess if you had -2 boxes with -3 beads in each of them. The best thing is to say that we are playing the game of algebra according to some rules and that $(-2)\times(-3)=6$ according to those rules.

Now, if challenged to solve $3x=2$, for example, you have to invent a new number, and this brings us to the operation of division ($\div$), an operation which increases the number system by all fractions, positive or negative, and this gives us the set of all *rational* numbers, with rules for manipulating them with the signs $+$, $-$, $\times$, $\div$ without passing out of that set. There is one forbidden act, division by zero; it appears profitless to try to insist that the equation $0\times x=1$ has a solution.

It is curious that this should be so, for we would never have got beyond the positive integers had we not thought of insisting that certain equations *must* have solutions, and now we shall go on doing that sort of thing. Consider the equation $x^2=2$. It has no solution belonging to the set of rational numbers and our first reaction may well be to reject this equation as meaningless. But here again it is wise to insist that a solution exists, even though it involves creating a new number.

It would of course be rather futile merely to insist that the equation $x^2=2$, and other similar equations, have solutions. One has to go on to say how they are to be manipulated with $+$, $-$, $\times$, $\div$, following the general rules of algebra, the same rules as are in force for the rationals. This can be done, not only for numbers invented to satisfy algebraic equations, but also for numbers like π, defined by limiting processes. Thus we create the *irrational* numbers, completing the *real* number system.

Having won control over an M-world inhabited by the real numbers and ruled by the laws of algebra, we might be content to rest. But there is no harm in going a little further, raising the standard of rebellion against once-accepted prejudices.

Consider the equation $x^2+1=0$. Since the square of every real number is positive (except the square of zero, which is zero), it is clear that this equation has no solution, unless we enlarge the number system in some way. This enlargement gives us the *imaginary* number $\mathrm{i}=\sqrt{-1}$ and *complex* numbers, each the sum of a real and imaginary part (say $a+b\mathrm{i}$ where a and b are real). Once again the question arises whether such numbers can be manipulated in accordance with the rules of algebra, and it is found that they can be so manipulated. Indeed the manipulation is deceptively easy: you can carry out a calculation with complex numbers just as if they were real rational numbers, substituting -1 whenever you come across the square of i.

The word 'imaginary' has one advantage: it points to the fact that we are definitely in an algebraic M-world. But it is a misnomer if it suggests that we left the R-world only when we tried to solve $x^2+1=0$. We left the R-world much earlier – in my opinion when we thought about integers deprived of names of things – in the opinion of Kronecker when we thought about negative numbers and fractions. Actually, one can discuss imaginary numbers without the symbol i or an equivalent symbol, for a complex number may be thought of as a pair of real numbers.

It is possible to think of the development of the number system in terms of a political analogue. The numbers are the inhabitants of a country. The population is infinite, but no matter. This population is increased by admitting immigrants to citizenship (the immigrants are far more numerous than the original inhabitants, uncountable in fact). The country is ruled by simple but rigorous laws, embodied in a constitution (the algebraic rules), and every new citizen has to take an oath to obey those laws. All goes well with this country, even though it has increased its citizenship to include the complex numbers.

But there are those who would tamper with the constitution. They would alter the commutative law of multiplication which states that $x \times y = y \times x$. Once that point is yielded, there is a flood of new immigration, notably quaternions and matrices, which behave in an exemplary manner as regards the other laws of algebra, but are frankly non-commutative when it comes to multiplication.

I do not find it necessary to end this chapter with an apology as I did the last one. It may well be that I have stressed the wrong points, telling you what you know already or omitting explanations where they were required. But the subject matter is a rock to build on, a firm mathematical rock not cracked by the intrusion of physical ideas which may change from age to age. Most emphatically this is an M-world, the M-world of algebra, austerely abstract in its upper regions, but with its roots in the R-world, the world of small children counting their fingers and toes.

CHAPTER 4

Variables, operators, functions

It is all very well to insist that algebra is utterly reliable while geometry must be treated with circumspection, but we are only human and logical reasoning is not enough. We need the flash of intuition, and that is supplied more readily by geometry than by algebra, at least to most of us. Algebra is the wife, utterly reliable but really too logical, and geometry the girl who enables us to look at things from a new angle (to borrow one of her own terms).

We have had the real number system before us and I compared it to a country with uncountably many inhabitants. How is the king of that country to think about them all? It appears inconceivable that he should be able to do this, but he can, by making use of a simple geometrical device. On a sheet of paper he draws a straight line, puts zero at the middle and measures off 1, 2, 3... to the right and −1, −2, −3,... to the left, as in fig. 4.1.

This representation has two obvious defects, but they are not serious. The line on paper cannot go to infinity in both directions (or even in one), but that defect can be made good by imagination – stick on more pieces of paper

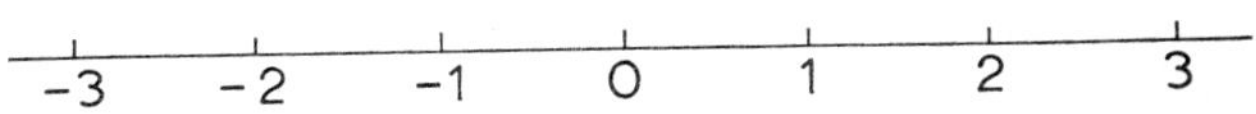

Fig. 4.1

Real numbers as points on a line

as required. Secondly, it is not possible to mark in all numbers, even between 0 and 1. But that again can be remedied by imagination, for any specified number can be marked in as required. The straight line is a powerful tool to help the intuition to deal with real numbers. Complex numbers are represented by points in a plane, but that does not concern us here.

The straight-line device is particularly powerful when it comes to the concept of a *variable* x which may take any value from minus infinity to plus infinity, or any value in a shorter range, say (0, 1). For this concept is represented by the whole line, or the prescribed bit of it.

Since the mind tends to seek particulars and shun generalities, I believe that this concept of a variable is not at all an easy one. Any statement made about a variable x must be true for *all values* of x in the specified range. It is not enough that it holds for this value or that – it must hold for all if it is to be a statement about the variable. For example, if x lies between 0 and 1, then x is greater than x^2.

Having conceived a variable x, the next thing is a *function* of x. We say that a variable y is a function of x if any particular value of x defines a particular value of y, and we write $y=f(x)$. The word *function* is unfortunate because it is a word in common use with other meanings, and for some purposes I prefer the word *operator*, but with a shade of difference in the precise meaning. (In Latin, *fungor* means *I perform*.)

Thus, if $y=x^2$, then y is a function of x. But at the same time we may say that a certain operation is applied to x, turning it into x^2; this operator says 'Square it!' Thus in general we have a choice of two notations: first, we can write $y=f(x)$, the functional notation, and secondly we can write $y=fx$ where f is an operator instructing us to do something to x. The notation fx does not, of course, mean that f is to be multiplied by x, for you cannot multiply a number by an operator. The difference between the two notations is psychological rather than real.

We may harness geometry again to help our intuition, drawing a graph of the function $y=f(x)$. This is a most useful thing to do because, by glancing at the graph, we can instantaneously appreciate the broad features of the functional relationship between x and y. But it has certain shortcomings, as we shall see on p. 41.

I propose another intuitive scheme, less satisfactory in some ways but better in others. Imagine a box with two rods sticking out of it, one labelled x and the other y (fig. 4.2). The value of the variable x is represented by the length of the rod x which sticks out of the box, and similarly for the variable y. If those two variables were independent, you could pull out the two rods independently to any desired extent. But if $y=f(x)$ (or equivalently $y=fx$), that is no longer the case. When you pull x out, y responds in a definite manner. You are to imagine some mechanism inside the box which connects the rods. This mechanism (whatever it may be, and we don't seek to specify it), or the box containing the mechanism, may be called a *function-box*, or *operator-box*. It is at worst a playful idea which does no harm. It has however a certain psychological advantage over a graph. If I ask you to think of all possible functions or operators, or a great variety of them, you may get dizzy trying to think of a vast variety of graphs drawn on a sheet of paper. It is, to me at any rate, easier to think of a vast number of boxes looking very much the same on the outside, but with all sorts of different mechanisms inside them.

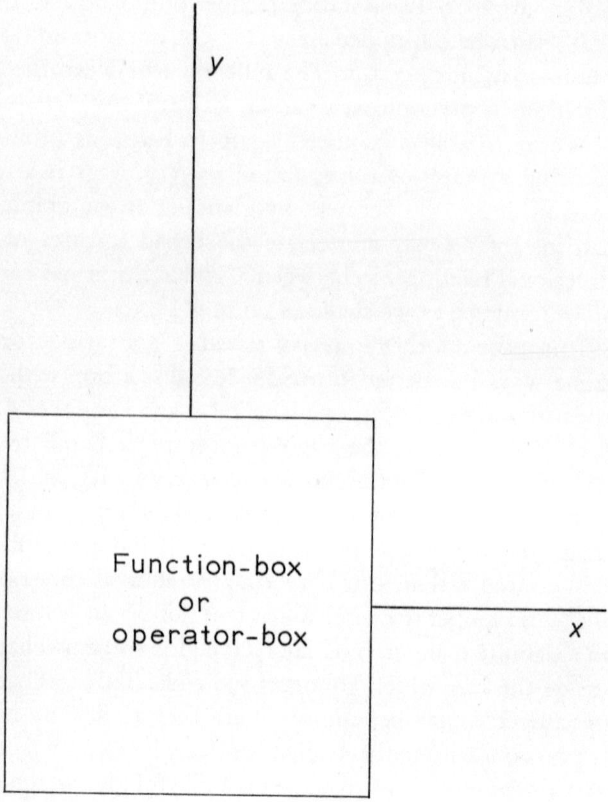

Fig. 4.2

Function-box (or operator-box) for one function of one variable

To show the function-box to better advantage, I think of four independent variables x_1, x_2, x_3, x_4 instead of a single variable x. I choose four, since that is the number suited to relativity, although what follows has nothing to do with relativity specifically. Let $y_1, y_2, \ldots y_{10}$ be ten functions of the x's (again I choose a number appropriate to relativity). This means that the values of the four x's determine the values of the ten y's. We might write out ten equations to express this, the first being

$$y_1 = f_1(x_1, x_2, x_3, x_4)\ .$$

But this is tedious. Suppose we let the letter x stand for the set of four x's and y for the set of ten y's. Then we can express the functional relationship in a single equation $y = f(x)$ or, operationally, $y = fx$.

Suppose we get muddled and want to see the situation intuitively. Our first thought is a graph. When there was only one x and one y, we got on nicely with a sheet of paper, with one axis for x and one for y. But now we need four axes for x and ten axes for y, fourteen dimensions in all. Geometry in fourteen dimensions is certainly not an impossibility, provided it is an M-geometry, but it is not any help to our intuition. In fact, graphs fail to help us as soon as they demand more than three dimensions.

Provided you don't ask too much of it, the function-box idea works here as well as it did with one x and one y. You have merely to imagine four x-rods and ten y-rods, connected by some mechanism inside the box such that you can play as you like with the x-rods, but, if you set them in a certain way, the y-rods are automatically set in a certain way (fig. 4.3). The mechanism in the box now represents ten functions, but we may say that it represents only *one*

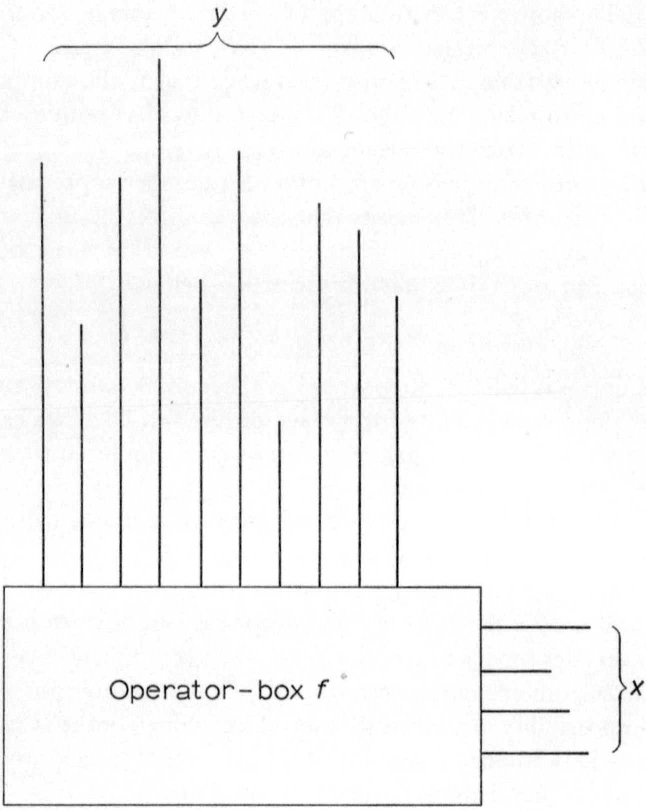

Fig. 4.3

Operator-box for four x's and ten y's which are functions of the x's ($y = fx$)

operator, namely the operator which changes the four-fold x into the ten-fold y.

Now what would happen if we were to link up two operator-boxes? An operator-box labelled A changes x into y, and an operator-box B changes y into z. Here x, y, and z stand for sets of variables, it doesn't matter how many there are in each set, but for definiteness let us take four x's and ten y's as before, and ten z's (again what we need in relativity). We have then

$$y = Ax, \qquad z = By ,$$

and hence $z = BAx$, if we replace y by Ax in the second equation. It appears then that x determines z, and so there must be some operator C that does the trick. If we write $z = Cx$, we seem to have

$$C = BA .$$

If these were numbers, not operators, we would recognise C as the product of the other two. This is a powerful suggestion, and we begin to wonder whether the laws of algebra hold for operators too. In general they do, but with one important exception, and that is the commutative law of multiplication.

Let us try it in the simple case where there is only one x, one y and one z. Let A mean 'Add 1 to it!' and B 'Square it!'. Then

$$ABx = A(x^2) = x^2 + 1 ,$$

$$BAx = B(x+1) = (x+1)^2 .$$

The two results differ, and so we conclude $AB \neq BA$ in general, although some operators do in fact commute.

Going back to the case considered above, it is usual to

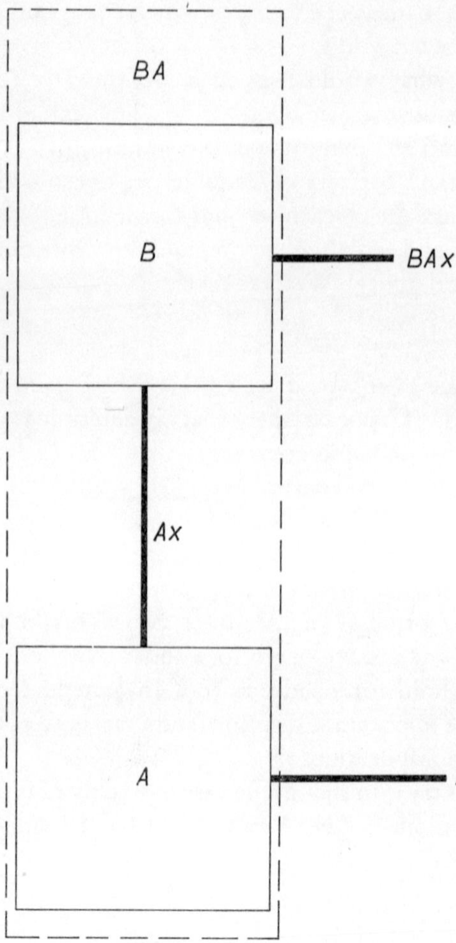

Fig. 4.4

The operator boxes A and B form an operator box BA

say that the y's are functions of the x's and the z's are functions of the y's. What are we to say of the relationship of the z's to the x's? They are functions of the x's, since the x's determine the y's and they in turn determine the z's. But that language does not express the complexity of the relationship, and it is customary to say that z is a functional or function of a function. But the operator notation says what should be said in a compact way.

As the matters about which I am now talking lie at the very heart of theoretical physics, it is by no means waste of time to discuss them somewhat further. Let me then remind you that x stands for a variable or, more generally, for a set of variables, perhaps four of them. The operator A gives instructions what to do with x, turning x into a new variable, or set of variables, called Ax. Then the operator B tells us what to do with Ax, turning it into the variable, or variables, called BAx. Note that in BAx we read from right to left, first x, then A and finally B.

It may help the imagination to make a picture of this in terms of operator-boxes. In fig. 4.4 we see operator-boxes labelled A and B. From A there protrude a set of rods x, shown as a single rod. These x-rods may be manipulated as you like, but their setting determines the setting of other rods, labelled Ax, connecting the box A with the box B. These Ax-rods operate a mechanism in box B, leading to a setting of other rods, labelled BAx, sticking out of B. If you are interested only in the connection between BAx and x, you may enclose the boxes A and B in a larger box called BA, in that order, *not* AB, because the operators in general don't commute.

What I am trying to bring out is the possibility, indeed the desirability, of thinking of these operators as inhabitants of the M-world, in spite of the fact that they are not numbers or variables, but merely orders, instructions to be carried out. This abstraction is given a sort of concreteness by thinking of them as boxes without inquiry about the mechanism in the boxes. If the boxes merely confuse you, then forget about them, and perhaps think up another way of helping your imagination to grasp these abstractions.

Some of the most important equations in physics are of the form

$$BAx=0 .$$

So let us have a look at this equation. It contains three things. There is the variable x, and I shall use the singular for it although it may actually be a set of variables. There are the two operators, A and B.

As we saw in talking about the number system, an equation is a challenge, a challenge to find something to satisfy it. Let us then start with the idea that the operators A and B are given (they convey known instructions) and that x is to be found. In terms of fig. 4.4, we are to find the setting of the x-rod which produces a zero setting of the BAx-rod. There may or may not be such a solution, but at any rate it is reasonable to try to find one. It is a question of finding some particular value for the variable x, not unlike the question of finding x to satisfy a given quadratic equation. Such a problem may often arise, but it is not that problem I am concerned with.

Suppose now that the operator B is given but A is not, nor is x given. The problem is to find the operator A. In terms of fig. 4.4, this means that the box B is given (it conveys known instructions) and x is free (you can pull the x-rod in and out as you please), and the problem is to find an A-box such that the BAx-rod remains at zero no matter what you do to the x-rod. You are not looking for a number – you are looking for instructions. This is indeed a central problem, although it may not be immediately recognised in this form.

Let me illustrate by some examples.

A body of mass m is moving freely under no forces, being in position y at time t. Its equation of motion reads

$$mD^2y=0\,,$$

where D stands for d/dt. Here the operator B is mD^2 (it says 'differentiate twice and multiply by m') and the operator A is the operator that turns t into y. Thus the equation of motion reads

$$BAt=0,$$

just what we had before except that it seems natural to use t instead of x for the variable. Note that B is given and what we seek is the operator A, with $y=At$. This is the equation with which Newton started modern physics, and the problem is to find A. Now we know what the solution is:

$$y=at+b,$$

where a and b are constants which depend on how we start the body off. Thus the operator A conveys the instruction 'multiply by a constant and add another constant'. The job is done.

Now consider a violin string vibrating. If y is the sideways displacement of the string, s the distance along the string and t the time, the equation of motion of the string reads

$$c^2D_s{}^2y-D_t{}^2y=0\,,$$

where c is a constant depending on the tension and density of the string, and D_s and D_t indicate partial differentiation with respect to s and t. Here the operator B is

$$B = c^2 D_s^2 - D_t^2,$$

and conveys the instruction 'differentiate twice with respect to s, multiply the result by c^2, and subtract from this the result of differentiating twice with respect to t'. There are two variables, s and t; let us use x to denote the pair. Then y is some function of x, say $y = Ax$. Now our equation reads

$$BAx = 0\,,$$

and the problem is to find the operator A. The solution is known to be

$$y = f(s+ct) + g(s-ct)\,,$$

where the functions f and g are arbitrary. Thus the operator A which we were seeking is that operator which tells us to do just that to s and t, that is, to x, which represents the pair of them.

One more example, which, with the vibrating string, may be regarded as the progenitor of modern field theories. Now let x stand for the three variables (x_1, x_2, x_3), coordinates in space, and let V be the potential (according to Newtonian theory) of the gravitational attraction of whatever masses there may be. Then V is some function of position, and we may write $V = Ax$ where A is some operator which is to be found. Finally let B be the Laplacian operator, telling us to differentiate twice with respect to each of the three coordinates and then add the results together. Laplace's equation then reads $BAx = 0$ or in more usual notation $\Delta Vx = 0$.

Let us not bother about the solutions of Laplace's equation. The point is that in all these cases there is a variable x and two operators, A and B, with B standing out in front, so that we may call it the *leading operator*. In all these cases the leading operator is given and the other one is not given – it is to be found – and a vast domain in physics poses two questions: (i) What is the appropriate leading operator? and (ii) Given the leading operator, how is the other operator to be found?

In modern times it has become customary to speak of theoretical physics and mathematical physics and to make a rather vague distinction between them. But I think it would be correct to say that theoretical physicists are supposed to be concerned with Question (i) and mathematical physicists with Question (ii). You might say that the former are concerned with discovering laws of nature and the latter with working out the consequences of those laws, for the appropriate leading operators are to be regarded as enshrining laws of nature. But such a clear-cut distinction is not realistic, because a law of nature is of little interest if you cannot develop its consequences, and so a theoretical physicist is unlikely to spend all his time trying to discover laws of nature. He, too, spends a lot of time trying to find A when B is given.

To be a little more general, one might say that a vast domain of physics is concerned with equations of the form

$$BAx = Cx\,,$$

in which x is the variable, B and C are known operators, and A is to be found. These are usually partial differential equations, but not necessarily so. But here again the leading operator B lies at the heart of the matter, and the discovery of an appropriate B is a mile-stone in the history of physics. The three B's I have talked about above were of supreme importance. In modern times the most conspicuous are the leading operators (equivalently, partial differential equations) invented by Einstein, Schrödinger and Dirac. The last two shared a Nobel prize for their inventions. Einstein won a Prize, but for other work. Later on, I shall be talking about Einstein's leading operator, but in general terms and without worrying you with details about the mechanism in his operator-box.

Let me remind you that though I have brought you back into the R-world by asking you to think about rods and boxes, that was only to stimulate the imagination. The concepts involved belong, needless to say, to the M-world of theoretical physics, no matter how well, or how badly, they predict the results of R-world experiments.

CHAPTER 5

Pontificating about events

Now had they brought the work by wondrous Art
Pontifical, a ridge of pendent Rock
Over the vast Abyss, following the track
Of Satan, ...

I shall not pontificate about events in the manner of the Pope speaking *ex cathedra*. Copying Milton's angels, I aim to bridge the abyss between Newton and Einstein by means of the concept of an *event*. For that word is in common use and has been employed in physics for a long time with a meaning very like the common one, but shorn of some dramatic aura (It was quite an event!), and it can be taken over into relativity where it plays a fundamental part. Of all concepts it is the most basic, and one may say that the event is to relativity what the point is to geometry.

But one thing must be made clear. A Newtonian physicist, challenged to define an event, might say that it is essentially a set of four numbers, namely, three coordinates (x, y, z) fixing the position of the event and a fourth (t) fixing its instant in time. That is an excellent way to put it if you are building up the M_1-world of Newton, but it will not do at all for the M_2-world of Einstein. It is true that the concept of event is carried over into the M_2-world, but much more cautiously, without reference to coordinates in space or an instant in time, for neither space alone nor time alone is an acceptable concept. If we accept those concepts at this stage, we stumble and fall before the race has begun.

Up to now I have been talking about various things. I have set up straw men to illustrate this or that and then I have knocked them down, as in the case of geometry, so that you may feel in doubt as to where you are expected to stand at this critical point where we are beginning to go relativistic. M-geometry must be treated with suspicion, not because it is imperfect in itself but because it has no place in the M-world of relativity until brought in with due circumspection, and then only in special relativity, and that is still far away. On the other hand the M-world of algebra (in the wide sense) is completely reliable and is to be regarded as always available to help us as we may need it. In fact, perhaps the best way to deal with the concept of an event is to say that it is determined by four numbers, say x_1, x_2, x_3, x_4, and then talk about the matter.

The first thing to say is that unless the creator of an M-world is a bit of a punster, his M-world will be of no use at all to physicists interested in the R-world. It will be a closed domain with no exit to reality. We must pun by using the same word for a concept in the M-world and something in the R-world. Now in the R-world the striking of a match is an event, and we pun by using this one word *event* for this and for a set of four numbers. But that is not good enough. We want to know how a person would go about assigning four numbers to the striking of a match, not with absolute precision (nothing in the R-world has that), but with reasonable accuracy.

To do this I must ask you to accept, not only the unique and mysterious R-world and the M-worlds which are as it were caricatures of it, but also an F-world, a world of fantasy. It is indeed essential in scientific thinking. What would happen if...? But nothing happens, except in the imagination. These are thought-experiments, much used in modern physics, and indeed probably throughout the ages.

So please imagine yourself sitting inside a room of some sort equipped with pencil and paper and a piece of apparatus which I shall call a *pseudo-clock*. Later on I shall give you a real clock, a clock which in some sense goes properly, but at this early stage nothing but confusion would result if questions were raised about the proper way to make clocks. Your pseudo-clock is very primitive. It has a hand that travels round a dial with numbers on it, and always travels the same way, never stopping.* The question whether it goes fast or slow does not arise.

Your duty is to note the reading of the pseudo-clock immediately an event is drawn to your attention. This is easy and direct for events occurring in your room, but most events happen elsewhere and you would know nothing about them unless you were told by some sort of messenger. That messenger might be a light-ray or radio wave, or it might be a boy on a bicycle. It is not your business to inquire, at least not now; your duty is to note

* It might be better to think of something like a gasmeter, with a row of connected dials, or like an electricity meter with digital readings. The essential idea is a device which gives numerical readings, steadily increasing, and ranging, in principle, from minus infinity to plus infinity.

down the reading on your pseudo-clock when you receive the message about the event.

If you like, you may say that you make the note *when you receive* the message. It would be most improper to say that you made the note *when the event occurred* or to inquire how fast the messenger travelled. If you do make such inquiries, I can merely say that, at this stage of the game, they cannot be entertained since they involve concepts of *simultaneity* and *speed*, and the first of those concepts will be rejected as unrelativistic while the second will call for definition and discussion at a much later stage.

Any event, then, which comes to your attention receives a number, say x, this being the reading of your pseudo-clock. It seems reasonable to expect that, if you are informed about a lot of events, many of them will receive the same value of x, or values so close as to be indistinguishable. In fact, if we call x a *coordinate* of an event, one coordinate does not suffice to distinguish one event from another.

It is convenient to have a name for the totality of all possible events. We might call it the *universe*, but *space-time* is more commonly used, and I shall use it, but with the important reservation that the hyphen is *not* to be removed except with much discussion and precaution – we are not to find ourselves talking about space and time, separately, as if we knew what those words mean.

Since one coordinate does not suffice to describe an event, we may say that space-time is not one-dimensional – merely another way of saying the same thing. Space-time must have a higher dimensionality, two, three, four, or even higher.

This brings us to the most fundamental hypothesis ever made – *space-time is four-dimensional.* There is nothing startling or new about this, since no doubt men thought in this way for a long time back, and a hawk poised for a strike must be aware instinctively of this four-dimensionality. Nor is there any deviation here from Newtonian ideas. All I assert is that the dimensionality is 4, and that if we write $4=3+1$ we are merely recording an arithmetical truth and not asserting that space-time may be separated into 3-dimensional space and 1-dimensional time.

Admitting, then, that space-time is four-dimensional, it is clear that you in your room with your pseudo-clock are not enough to describe events. Instead of one reading x, there must be four (x_1, x_2, x_3, x_4), and so you need four observers in four rooms, each with a pseudo-clock, and each recording in his notebook the reading of his pseudo-clock when he is informed of the event. By this means four coordinates are attached to each event.

Physicists and astronomers are in the habit of attaching four coordinates to events, but they do not use four observers in rooms with pseudo-clocks, and there is no reason why they should. It is the *fourness* of the operation that is important, not the way it is carried out. Once that is established, I can afford to play with the idea, introducing perhaps some Newtonian concepts without the fear that you will be contaminated by them.

Suppose you are in a vast desert in black night in which you can see nothing but the luminous face of your wrist-watch. Your blood curdles on hearing the roar of a lion, followed immediately by a second roar. You happen to be deaf in one ear and so have no sense of direction in regard to sound, and so you cannot tell whether the two roars came from the same direction or from different directions. The only thing you can do is to look at your wrist-watch and note its reading. Incidentally, I may note that your wrist-watch is in very bad repair – it keeps going, that is all you can say for it.

Now what you would like to know is whether there is only one lion or two. It might be that one lion made the two roars, in which case that might be regarded as a single event. But if two lions were responsible, they might not have been near each other, and then there would be two events. As you have only one cartridge left, you would like to know whether it was one event or two. But you have only one watch-reading, and that is not enough.

However there are others in the desert. They too have heard the roars, they too have noted the readings of their wrist-watches, and they too may be wondering whether it was one event or two. The best thing to do is to come together and compare notes. So, clutching your notebook with the figure x_1 entered in it, you go to look for the others and consult with them.

In the notebook of the first man you meet there is only one figure x_2. It seems probable that, if there were two events, not one, then he would have written down two numbers, not one. But that does not settle the matter.

You must seek another man. If his notebook contains two readings, then surely there were two events, because a single event can give only one reading. But if his notebook shows only one reading, what then? If I were you I would conclude that there was only one event, hence only one lion, and I would go to shoot him with my single cartridge, with some misgivings at the thought of silent lions or perhaps two lions roaring in juxtaposition.

But why should you be satisfied with three observers if space-time is four-dimensional? There is a catch in it.

Lions roaring in a desert form, or may be deemed to form, a continuum of three dimensions. If they flew, then you would need four observers.

I am prepared to repeat ad nauseam that the real R-world is so complicated that one must hesitate to make dogmatic statements about it, and so I would brush aside the question whether it is four-dimensional or not. It is rather a question to be asked about any M-world which we construct as a model of the R-world, and Newton and Einstein agreed that four-dimensionality was to be built into their models. Still, one is permitted to wonder, and it may not be without profit to devote another chapter to the matter.

CHAPTER 6

The Queen and the Captain of the Guard

The King of Pseudosylvania had married a wife considerably younger than himself, and, as often happens in such cases, he suspected that his Queen was unfaithful to him, the object of her affections being the young and handsome Captain of the Guard. Racked with doubt as to the validity of his suspicions, he summoned the President of the Royal Academy of Sciences and asked for his advice.

'Well,' said the President after thinking the matter over, 'Nothing improper can occur unless their world lines meet.'

'World lines!' snorted the King. 'Never heard of them. Explain yourself, man!'

So the President asked for a blackboard. When this had been brought and the attendants had withdrawn, he marked four small crosses on it. 'Those,' he said, 'are four events in the life of her Majesty the Queen.'

'Nonsense!' cried the King. 'She hasn't had any babies yet. And,' he added pathetically, 'I don't think its my fault.'

'Sire,' retorted the President, 'I am not an obstetrician but a physicist, and such matters are quite out of my line.

Those marks merely represent events in the most undramatic sense of the word. This mark, for example, might represent her Majesty having breakfast this morning.'

'But why pick out four?' asked the King.

'The number four was, on this occasion, a mere accident,' replied the President. 'I might have made two marks or I might have made twenty. The number of events in her Majesty's life from birth to death (may she be long preserved!) may be regarded as infinite. One event follows another in continuous succession. So, if your Majesty will permit me, I shall amend my diagram.'

Whereupon, going to the blackboard, he erased the four crosses and drew instead a curve extending from the bottom of the blackboard to the top. 'There,' he remarked with some complacency, 'is her Majesty's world line. At the bottom I write B for birth. At the top I put D for death, but with a note of interrogation. As I indicated, I am not a medical man, but with the new drugs and transplantations, her Majesty might live for ever.'

The King looked grim. 'If I find her guilty,' he muttered, 'I would not give her a high life-expectation.'

'To proceed,' went on the President, 'I put an arrow on the curve, pointing upward. The curve represents her Majesty's world line and this is the arrow of time, pointing from her past into her future.'

'Surely that arrow is unnecessary,' said the King. 'How could she die before she was born? Ah, yes, of course, she might have died in her mother's womb. But then she didn't. If she had, we would be spared all this bruhaha.'

'Your Majesty is right in drawing attention to redundancy,' said the President. 'Redundant information is good when it confirms what we already believe to be true, but a great source of confusion if it conflicts. A king must realise that it may be wise to have a spy to spy on a spy.'

'Tush!' cried the King. 'I do not dirty my hands with spies. That is the business of my Minister of Information, or Misinformation as I sometimes call him in jest. But go on. No possible exception can be taken to the Queen's world line as you have drawn it. Where does the Captain come in?'

The President drew another curve distinct from the first, and sketched in a crown and a sword to show to whom they respectively belonged.

'Excellent!' cried the King, rubbing his hands in glee. 'You find her innocent, then? With world lines as far apart as that, I cannot see that any impropriety can occur. But what about their thoughts? Can you put them in too?'

The President coughed in some slight embarrassment. 'We must not be hasty,' he said slowly. 'This diagram is not to be regarded as a statement of fact, but merely as a guide for our reasoning, a guide for the determination of facts. Look,' he went on, and he altered the Captain's world line so that it made contact with the Queen's. 'Or look again,' and he altered it further so that it lay along the Queen's line for several inches.

'Foul bitch!' shouted the King. 'She has betrayed me! They shall both suffer the extreme penalty!'

Seizing the duster, he rushed to the blackboard and rubbed out the two lines just above the fatal contact. Then he sank down into his chair with his head in his hands in an agony of thought, while the President stood by waiting silently. Presently the King got up and went slowly to the blackboard where he made some careful alterations. He erased the Captain's world line before it came into contact with the Queen's line, and then very carefully restored the latter to its complete original form.

'That's better,' he said smiling. 'I liquidate the Captain, see. Nothing happens. The Queen is innocent, and all is well. But,' he added, 'where do I come in? And you? And everyone else?'

Reaching for the chalk, he was about to draw more world lines, but the President restrained him. 'The situation is complicated enough as it is,' he said. 'If you try to show everything in a diagram, it becomes useless. But, please your Majesty, pull yourself together. You cannot alter facts by rubbing things off a blackboard. And, moreover, you cannot execute a man for a crime he has not committed.'

'But perhaps he meant to commit it!' cried the King. 'Perhaps he has guilty thoughts. Yes, I ask you again where thoughts are in your diagram.'

'Until we know more about the physiology of the brain,' answered the President, 'we can hardly attempt to depict thoughts or emotions in such a diagram. If I may suggest it, with all respect, it is wiser not to bother too much about what people think. They hardly know themselves. But,' he went on, brightening, 'messages or signals are a different matter.'

Going to the blackboard, he eliminated the Captain's death by completing his world line as it stood originally, clear of the Queen's. Then he drew a line connecting the two world lines, sloping up from the Queen's to the Captain's, with an arrow pointing in that sense.

'Here,' he explained, 'is a message from the Queen to the Captain.'

'What does the message say?' asked the King anxiously.

'I have no idea,' replied the President. 'The situation is purely hypothetical. The message might go the other way.' He drew a line diagonally upward from the Captain's world line to the Queen's.

'Why do you draw those lines slanting upward?' asked the King. 'Why don't you draw them straight across?'

'Because it is the proper thing to do,' answered the President a little stiffly. 'Suppose there is an exchange of notes...'

'Do you have to torture me?' cried the King.

'Suppose there is an exchange of notes,' continued the President inexorably. 'The Captain sends a message to the Queen and she sends a message in reply...'

'That is not quite as bad as it might be,' the King conceded. 'Her reply might repel his advances. I hope that it is so.'

'That is not the question,' said the President. 'The Captain sends a message to the Queen and she sends a message in reply...'

'You said that already,' growled the King.

'I must ask your Majesty not to interrupt. The Captain sends a message to the Queen and she sends a message in reply. I suppose that she sends it immediately.'

'She *is* very impulsive,' interjected the King.

'If you will allow me to make the point, your Majesty,' went on the President, 'it is that the Captain has to *wait* for the reply.'

'Of course he has to wait – I wish he would wait for all eternity!' shouted the King. 'Of course a messenger takes time to come and go. You don't have to tell me that.'

'If your Majesty would condescend to inspect a diagram, it might help,' said the President patiently. Leaving the two world lines untouched, he erased the diagonal lines he had previously drawn to represent messages. Then, starting from a point on the Captain's world line, he drew a line sloping *downwards* to the Queen's and from there another line, also sloping downwards, back to the Captain's world line. He marked in more strongly the arrows of time, pointing upwards on the two world lines, and then inserted arrows on the sloping lines in the sense Captain to Queen and back to Captain.

'Now, your Majesty,' he said when this was done, 'I have made the message-lines slope down, not up. See what happens. Following the arrows on the sloping lines, we see that the Captain receives the reply before he sent the message.'

'But that is absurd!' cried the King.

'Of course it is,' said the President, 'and that is why I like to draw message-lines sloping up, not down, when you follow the sense indicated by the arrow.'

'I see your point,' said the King. 'But time is passing and we don't seem to have made any progress in finding out whether the Queen is guilty or not.'

'I am at your Majesty's service,' said the President, bowing, 'but I would like to add a remark to what I was saying about the time-lag which must occur between the sending of a message and receiving a reply to it. It is perhaps wrong to say that there is always a time-lag. If, for example, her Majesty and the Captain were lying in each other's arms, the time-lag might be equated to zero.'

'Stop it! Stop it! Stop!' groaned the King, putting his fingers in his ears.

When he had recovered, the President started to outline plans for determining the innocence or guilt of the Queen.

'Your Majesty might engage the services of two detectives, one to tail the Queen and the other to tail the Captain. Call those detectives A and B. It would be the duty of A to make his world line coincide with the world line of her Majesty and the duty of B to make his coincide with that of the Captain. Then the world lines of the Queen and the Captain would meet if, and only if, the world lines of A and B met. If you provided two reserve detectives, C and D, to take over the job temporarily, you could receive reports from A and B (and in due course from C and D) as to whether the world lines had met. This is the method of coincidences and is thought highly of.'

The King looked at him quizzically. 'Do you mean,' he asked, 'that if her Majesty and the Captain got into bed together, then A and B (or C and D) would have to get in too?'

The President blushed. 'Your Majesty's mind is, if anything, too acute. It would probably suffice if the detectives looked through the keyhole.'

'Two detectives would need two keyholes,' sneered the King. 'Wouldn't one detective do?'

'Yes, indeed,' replied the President, taken aback. 'After all, why use two? It would be enough to have one to tail the Queen.'

'I am beginning to lose confidence in the judgment of the President of my Royal Academy of Sciences,' said the King a little bitterly. 'The dignity of her Majesty is not compatible with being tailed, as you call it. And if you put a detective to tail the Captain, he will be quickly run through by the Captain's sword. Have you any other idea of how to go about it?'

The President thought it over and then came out with a

plan. It involved, not one detective or two, but four special agents, each equipped with a clock and a notebook. The clock, he explained, need not be a valuable clock, provided it kept going and never went backwards. The King roared with laughter at the idea of a clock going backwards and this put him in the best of spirits.

These four agents, the President explained, would receive signals from the Queen and the Captain, and would note down the times when these signals were received.

'But why,' asked the King, 'do you think they would send signals, except possibly from the one to the other, that is, if there is guilt, and that is what we are to investigate?'

'Her Majesty chatters a lot,' began the President, but changed his mind and said 'Her Majesty has a ready flow of conversation, and the waves of sound carrying her conversation could be picked up at a considerable distance by delicate auditory apparatus. The Captain is inclined to jingle his spurs, and that too would make a signal. But even if they were silent, they would emit heat-waves which it would be possible to detect. If your Majesty is prepared to lay out a little money to equip these four agents with suitable apparatus, in addition to clocks and notebooks, the scheme is entirely feasible.'

'What scheme?' asked the King, yawning slightly. 'I have so far heard of no scheme, but a lot of nonsense about waves of sound and heat. Why not increase the absurdity by bringing in light-waves and radar?'

'Indeed we might,' replied the President thoughtfully. 'But the difficulty is to get the signals to go through matter. If her Majesty and the Captain were strolling in the Royal Park, the four special agents could actually see them and would need no more apparatus than their eyes. But if the couple were to lie down in a hollow, they would be lost to view and the agents could not make any further report until they emerged, unless you thought fit to supply delicate seismometers. Perhaps we might use neutrinos–they go through nearly anything.'

'Yes, yes, yes!' cried the King impatiently. 'Those are technical matters, and I leave them to you, but do not exhaust the Royal Purse. Get on with the scheme. What are these special agents to do after they have received the signals?'

'In order to make the survey adequate,' explained the President, 'you will actually need eight special agents, not four, so that there may be four on duty and four resting. Perhaps it would be better still to employ twelve, so that we would have four taking observations, four working out deductions from the observations, and four resting.'

'Those are details,' muttered the King. 'Keep going.'

The President continued. 'When the four agents come back after making observations, each has a notebook containing a column of numbers, increasing as you go down the page. They hand those notebooks to the four agents detailed to process them.'

'That is an American term,' interjected the King. 'I was never able to make out just what it means.'

'Your Majesty is most kind,' said the President humbly. 'It was a Freudian slip on my part. I used a word I did not properly understand in order to cover the fact that, as I suddenly realised subconsciously, I did not properly understand what came next. You must forgive me for misleading you a little. The notebooks of the four agents must contain more than mere numbers. They must contain remarks serving to identify the events observed. That is most important. The events observed by the four agents must be the same events and be recognised as the same events. Thus the numbers must be accompanied by little notes like "Queen breakfasting" or "Captain shaving" and so on. With this amendment I can proceed, if your Majesty will allow me.'

'Proceed,' ordered the King graciously.

'Perhaps,' suggested the President, 'it would help if I used the blackboard,' and he wiped it clean. 'I am going to act the part of the four agents who are to deal with the data presented in the four notebooks. The first step is to rule four columns on the blackboard and head them with the names of the four agents who made the observations.'

'I take it,' remarked the King, 'that the four agents now dealing with the observations are not the four agents who made them.'

'Correct, your Majesty,' said the President. 'But permit me to say that it is a trivial detail. They might be the same men, but I thought it only right to let the four observers retire for a rest. But there is no reason why they should not deal with the data they collected. Indeed it might be better that way, if it would not be throwing too great a burden on them, for they might be able to supplement the terse remarks which identify the events by less formal comments, little imaginative quirks helpful to the memory, for it is of prime importance that the events be identified. Otherwise the whole investigation will be fruitless. Let me then change what I said and assert that I now represent the four observers who have come in and are standing in front of the blackboard as I am standing here. Is that clear?'

'Admirably,' said the King, smiling, 'and that reminds

me of an old story about soup. My father used to tell it, and you might like to hear it.'

'At another time, if it please your Majesty,' said the President quickly. 'I suggest that we continue with the four observers whom I represent, and so I proceed to examine the four notebooks. I decide to pick out some items, such as "Queen breakfasting" and "Captain shaving", and look for them in the four notebooks. In each notebook there is a number opposite such a note, and I write them up like this:

Event	Name of observer and his entry			
	Sam	Bob	Bill	Jim
Queen breakfasting	23	736	73	1134
Captain shaving	436	196	32	226

In general we have a named event followed by four numbers which may be denoted by x_1, x_2, x_3, x_4.'

'I never thought I would see algebra involved with adultery,' said the King pensively.

'It may not be adultery,' said the President kindly, 'and anyway this is hardly algebra. Those letters merely stand for numbers written on a blackboard.'

The King gazed at the blackboard. 'Tell me,' he asked, 'which came first – the Queen's breakfast or the Captain's shave?'

The President was taken aback. 'Why,' he stammered, 'I have no idea. I made up those eight numbers at random to show you the principle of the thing, and they may make nonsense. It is true that, if we look at Sam's figures, the higher number is attached to the Captain's shave, and that means that Sam received the signal from the shave (a flash of a shaving mirror, perhaps) after he received the signal from the Queen's breakfast (maybe the crunch of toast). On the other hand, the two signals reached Bob in the reverse order, and this happens to be the same order as observed by Bill and Jim. If your Majesty wishes, I can look into the matter, and indeed I would be glad to do so, but it might take much time to elucidate and I suggest its postponement to a later date in order that I may proceed with the main theme, in which indeed the question you so graciously raise does not arise.'

The King glared at him stonily. 'If only you had provided the observers with decent clocks,' he said, 'the question would not arise at all. For then all four of them would have recorded the same time for the Queen's breakfast (perhaps it might be 23) and the same time for the Captain's shave (it might be 436), and then we would know that the Captain shaved after the Queen breakfasted, although for the life of me I cannot see what that would tell us about her fidelity or infidelity.'

The President drew himself up haughtily. 'Your Majesty,' he said coldly, 'is going too far. There are no *decent* clocks, clocks that record an absolute time with complete reliability. It is not the word *reliable* that is at issue but the word *absolute*.'

'I did not use the word *absolute*, I said *decent*,' muttered the King, surprised and somewhat cowed by the President's attitude.

'Your Majesty's meaning was clear from the context,' stormed the President, losing his temper. 'When you said *decent* you meant *absolute*, and this I find intolerable. Would you have us put the clock back (excuse me if I confuse the matter, but I am deeply moved) – would you have us return to the days of your illustrious grandfather (may he rest in peace!) and wallow in the misconceptions of an age barbarous in comparison with your enlightened reign? No, Sire, it would be asking too much of me. I would prefer to resign from the proud office I hold.' At this he burst into tears.

'Tut! Tut!' cried the King, taken aback by this display of emotion. 'I had no idea you took all this so seriously. Perhaps after all it would be simpler to get a detective to tail the Queen.'

The President pulled himself together. 'Nay,' he said, 'her Majesty must not be subjected to such indignity. The four observers provide a better way. If you will bear with me, I shall continue to describe the processing of the data.'

'Processing!' said the King smiling. 'But proceed.'

The President then told how the data would be used to draw graphs on two sheets of squared paper, one sheet for Sam and Bob and the other for Bill and Jim, but he made it clear that the four observers need not be grouped in this way. Sam might go with Bill, for example, and Bob with Jim. The results they were seeking, he said, would be invariant under this transformation, and the King let it pass as he was getting a bit tired.

'First,' said the President, 'we draw the Queen's graphs. Note that I use the plural. If our ordinary space had four dimensions instead of three, then one graph would suffice – it would be a curve in four dimensions. But that is not the case, and we have to break down the number 4 into 3+1 or 2+2. But to use the former, we would have to make a 3-dimensional graph with a piece of wire, and the 2+2 choice is easier, because all we need is two sheets of squared paper, each of them 2-dimensional. The event with the four co-ordinates (x_1, x_2, x_3, x_4) appears as two points, one point on one sheet of paper with coordinates (x_1, x_2) and the second sheet with coordinates (x_3, x_4).'

'The Queen's breakfast, then,' interposed the King, 'will appear as two points, one on each sheet, and the Captain's shave will likewise appear as two points, one on each sheet. Although the numbers you have put on the board indicate that it is not so, I would be glad if you would tell me what you would deduce if the two points representing the Captain's shave fell right on top of the two representing the Queen's breakfast.'

'I would deduce,' replied the President, 'that the Captain was shaving at her Majesty's breakfast table or that her Majesty was taking her breakfast in the Captain's bathroom. But indeed I am overreaching myself a little – it might all be happening in a hotel room. Such intimacy, though likely to make scandal spread, is not, if I may say so, sufficiently strong evidence of guilt to justify extreme punishment.'

'I would thank you,' snorted the King, 'to keep your ancillary remarks to yourself. I am trying to understand your scheme. Is it correct to say that in due course there will appear *two* curves which jointly represent the history or world line of the Queen, and likewise *two* curves for the Captain? Perhaps it would be well to draw the Queen's curves in purple and the Captain's in scarlet.'

The President bowed. 'Your Majesty has grasped the idea with quite extraordinary speed. I have hardly anything to add.'

'But I have a few more questions,' said the King. 'One is apt to get confused by too many words. We could of course get two sheets of squared paper and draw the curves we are talking about, but it is always dangerous to write things down, for someone seeing the graphs might conclude that they were a record of something that had actually happened, whereas in truth they would represent an imaginary situation which might or might not be true. Therefore I suggest that we form only mental images of those purple and scarlet curves. But to hold those images, we need to express ourselves very clearly.'

'What you say, if I may say so, your Majesty,' said the President, nodding his head, 'is indeed true. We must never confuse reality with things imagined, but on the other hand we can never understand reality unless we give play to the imagination.'

'Well,' continued the King, 'I propose to refer to one sheet of paper as the Sam–Bob sheet and to the other as the Bill–Jim sheet. When all the data have been inserted, I assume that the isolated points will be joined up to form smooth curves. Then on the Sam–Bob sheet there will be a purple curve and a scarlet curve, and likewise on the Bill–Jim sheet. Now suppose that each purple curve keeps away from the scarlet curve – suppose that there is no intersection. What do you deduce?'

'I deduce,' replied the President, 'the complete innocence of her Majesty, at least as far as the Captain is concerned. Much as I hesitate to say it, this non-intersection would be no proof of innocence in view of the existence of other world lines, not depicted here.'

'Stick to the point,' growled the King. 'I am concerned with the Captain only. Now a further question: if there is an intersection on one sheet, but not on the other, what then?'

'Innocence again,' replied the President.

'And if,' went on the King, 'there is an intersection on

each sheet, do you deduce guilt?'

'Hardly, your Majesty,' said the President. 'They might merely have passed close to each other in the street. Suspicion might be aroused by a succession of many intersections, but guilt would only be established if the purple and scarlet curves coalesced for some distance on both sheets.'

'Thank you for your trouble,' said the King, dismissing the President, after which he went in disguise to an agency and hired six special agents. His choice of the number *six* was due to confusion, parsimony and caution. The President had mentioned *twelve* agents and then had come down to *eight*, but in the end it seemed that *four* was the essential number. The King decided to do the processing himself for reasons of secrecy and economy, and to liquidate the agents as soon as they had handed in their notebooks. So he set out for the agency resolved to hire four agents and no more, but on the way he began to doubt whether the President really knew what he was talking about. Perhaps four would not suffice. Five would surely be better, but then what would become of the two sheets of squared paper? He settled for *six* since then it would be merely a question of using three sheets instead of two. So *six* was the number he hired.

Before the King had received any reports from the six agents, the private spies of the President of the Royal Academy of Sciences brought him word that the King had hired six special agents.

'Six'! cried the President to himself. 'What a strange number to choose! If the King uses them in two shifts of three each, they will be inadequate, for three coordinates are not enough to describe an event. They would be enough if the Queen and the Captain stayed on the ground, but not if they went upstairs, separately or together, and the whole investigation would be sheer waste of time.'

'But,' his soliloquy continued, 'if he uses all six agents at once, the information will be redundant, since four would suffice. Still, if the King chooses to waste money, that is his business.'

Nevertheless the President was nettled that his advice had not been taken and he decided to withhold any further assistance from the King. But just as he was lowering himself into his bath, after soaping himself all over, he sat up with a jerk. 'What a prejudiced fool I am!' he cried, and in a state of great excitement lowered himself again into the water, washed off the soap, got out and proceeded to discuss the matter with himself while sawing the towel across his back in front of the large mirror which he had in his bathroom, being a rather vain man.

'What a chance!' he exclaimed. 'What a wonderful opportunity to win a Pseudonobel Prize! Why be a slave to tradition, to Einstein, to Minkowski? Why should space-time have only four dimensions? Why indeed?'

Never in his experience had the matter been put to the test – here was the chance to do it. He decided to ask the King's permission to participate in the processing of the data supplied by the six special agents. The truth should reveal itself to the trained mind, the truth, that is, about the dimensionality of space-time, in comparison with which the truth about the Queen's adultery or non-adultery was a trivial thing indeed.

Being now completely dry, he put on a dressing-gown, went to his study, put three sheets of squared paper on the table together with a purple pencil and a scarlet one, and sat down with an air of extreme concentration.

He imagined the six special agents standing in a row in front of him, each with a notebook in his hand. Taking the notebooks from them, he wrote the numbers 1, 2, 3, 4, 5, 6 on the covers and dismissed the agents. Opening the notebooks, he looked through them for the notes identifying events, such as 'Queen breakfasting' and 'Captain shaving'. Corresponding to each event, thus identified, there were six numbers, one in each book, and these he wrote down in order. But as the six agents had not visited him and the six notebooks not actually there, and all this was going on in his imagination, he did not write down definite numbers, but the symbols

$$x_1, x_2, x_3, x_4, x_5, x_6 .$$

Taking one of the sheets of squared paper, he marked two axes of coordinates and labelled them x_1 and x_2. He did the same for the other two sheets, with labels x_3 and x_4 for the second sheet and labels x_5 and x_6 for the third.

'Now,' he said to himself, 'each event will appear as a point on each sheet, three points in all. When all the events are plotted in, they may be connected to form smooth curves, and there will be two curves on each sheet, one for the Queen and one for the Captain. We decided to draw the Queen's curves in purple and the Captain's in scarlet – the King suggested it and I did not reject the idea. Patrons must not be offended. But those colours are merely royalistic ritual. If boy meets girl, then girl meets boy, and the colours of their world lines have nothing to do with it. A pair of curves either meet or don't meet, and that is an end to it. I shall draw all curves in black, because it is simplest and least distracting, but I shall reserve the symbols

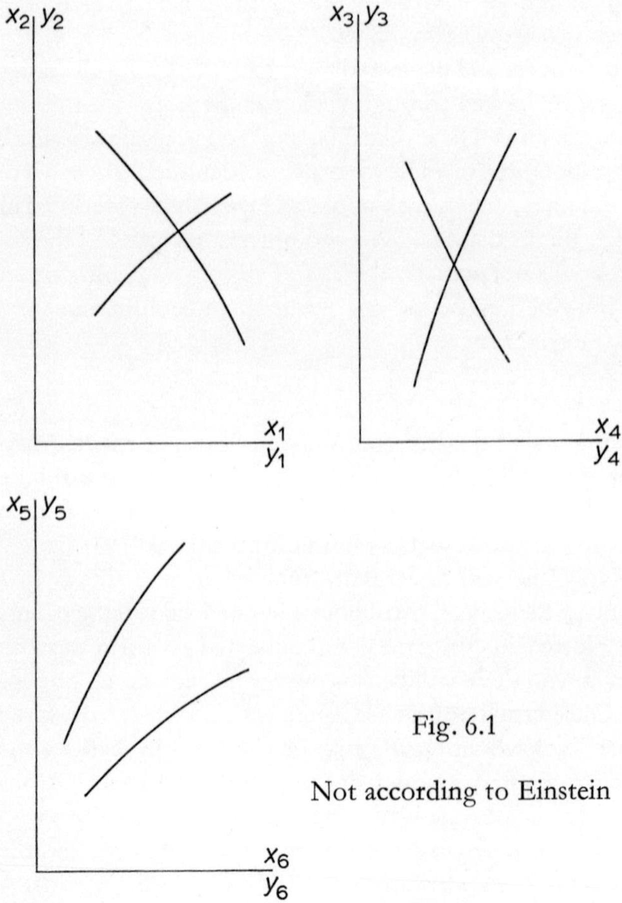

Fig. 6.1

Not according to Einstein

$$x_1, x_2, x_3, x_4, x_5, x_6$$

for the Queen's events and use

$$y_1, y_2, y_3, y_4, y_5, y_6$$

for the Captain's events. I shall of course plot (y_1, y_2) on the same sheet as (x_1, x_2), and likewise for the other pairs.'

The President sketched some curves at random on the three sheets, two curves on each sheet, and sat contemplating them. As it happened, the two curves on the first sheet intersected, the two curves on the second sheet intersected, but the two curves on the third sheet did not intersect (see fig. 6.1).

'Is that possible?' he asked himself. 'Have we got to the heart of the matter? Einstein on toast and me a Pseudo-nobel Laureate? Is it really as easy as that?'

He scribbled down some simple formulae. 'An intersection on the first sheet means that we have, at that intersection,

$$x_1 = y_1, \quad x_2 = y_2 .$$

An intersection on the second sheet means that, at that intersection,

$$x_3 = y_3, \quad x_4 = y_4 .$$

Now those two intersections represent, when taken together, an event common to the Queen and the Captain, and for it we have

$$x_1 = y_1, \quad x_2 = y_2, \quad x_3 = y_3, \quad x_4 = y_4 .$$

That is the way Einstein would have it, since he claims that four coordinates fix an event. But then, what about the curves on the third sheet – they do not intersect! But they ought to intersect. Because had we used the first and third sheets, instead of the first and second, we would not have found any event common to the Queen and the Captain – they would not have met. And it is contrary to elementary logical principles that they both met and did not meet.'

He rubbed his hands in glee, feeling that he had indeed got Einstein on toast, and that great fame would be his when his discovery was announced. But his enthusiasm was deflated by the reflection that this was only a thought-experiment – the curves he had drawn were merely random doodlings and not experimental facts. 'Still,' he thought, 'at least I now know what to look for.' Ringing for his valet, he told him to lay out his best suit and telephone the Palace for an appointment at his Majesty's earliest convenience.

When he got to the Palace, the King received him a little gruffly, embarrassed by the fact that he had hired six special agents instead of the number recommended by the President.

'Can I assist you Majesty with the processing of the data?' asked the President.

The King realised that, if he tried to do it by himself, he might get muddled and make a mistake disastrous to all the parties concerned. 'Well,' he said with some hesitation, 'your assistance would be much appreciated. Let us step into the Royal Library and look at the notebooks the special agents have just brought in.'

There they found six notebooks arranged in three piles.

'There seem to be six, not four,' said the President, smiling slightly.

'I recall that you recommended four,' said the King, 'but I thought it well to be on the safe side. Redundant information, as I think you remarked, does no harm, provided it presents no logical conflicts.'

The President did not reveal to the King that his interest had shifted from the Queen's sex-life to the dimensionality of space-time. He sat down beside the King, with three sheets of squared paper in front of them and pencils, purple and scarlet. But I cannot tell you what they found out. This tale belongs to the folklore of Pseudosylvania and has never been written down completely, but transmitted orally. No one has been able to remember it beyond this point at which the memory sinks back in exhaustion. I have not the least idea whether the Queen was adulterous or not, but I would hazard a guess that when the King and the President plotted out the purple and scarlet curves, they did not find the situation (fig. 6.1) which led the President to dream of fame. I believe that, if no mistakes were made, intersections on the first and second sheets inevitably were accompanied by an intersection on the third sheet. For otherwise it would appear that space-time is not four-dimensional, but at least five-dimensional, and a fact of such paramount importance to physics would have leaked out of Pseudosylvania to astound the world of science. But perhaps, before the President of the Academy had time to write up his results for publication, the King suspected him of harbouring deep and dangerous revolutionary thoughts, and had him liquidated just in case.

CHAPTER 7

Particles, world lines, and the arrow of time

In the R-world a particle is a very small thing, a speck of dust, a tiny grain of sand, or an object so small as to be invisible to the naked eye or even invisible under the strongest microscope. The operative word is *small*. When it comes to fitting the word *particle* into a physical M-world, no one who has got the Euclidean point under control should have any difficulty – a particle is a moving point.

The concept of a particle as a moving point is not foreign to Newtonian dynamics. It has long been customary to divide the subject into dynamics of a particle, dynamics of a rigid body, and dynamics of continua (such as fluids and elastic solids), but it is of interest that Newton himself did not create the concept of a point-particle. That concept was due to Boscovich* and caused much debate and con-

* It has been widely held that R.J. Boscovich (1711–1787) invented the concept of the point-particle (cf. L.L. Whyte, *Roger Joseph Boscovich*; Allen and Unwin, London, 1961), but C.A. Truesdell (*Essays in the History of Mechanics*; Springer-Verlag New York Inc., 1968, pp. 107, 282) attributes it to L. Euler (1707–1783).

troversy before it was finally given an honourable place in the M_1-world of Newton.

In passing to the M_2-world of relativity, we simply take over that concept of a particle, looking at it in a slightly different way. That is, having set up the concept of a four-dimensional space-time of events, we recognise a particle as a succession of events, a *world line*, this being in fact the history of the particle.

Some of what I have to say in the present chapter has already been introduced in the preceding chapter in a more fantastic setting, and there I described how a world line could be represented by a pair of graphs, each drawn on a sheet of squared paper. That was a sound plan, but inconvenient. We can use a more slap-dash representation in order to be able to think more quickly, and indeed that too has been anticipated. But let us look at the problem again.

The difficulty is, of course, the fact that space-time is four-dimensional, and it would be foolish to underestimate that difficulty. But the three-dimensionality of space is also a difficulty and mankind has learned how to overcome it. The engineer and the architect don't bother about space-time, but do have to think of building structures in three dimensions. Their usual procedure is to make diagrams on paper (plans and elevations), and less frequently to make three-dimensional models. The plans and elevations are projections on horizontal and vertical planes, and are analogous to the graphs discussed in the preceding chapter, sound but clumsy. If you want to get a general idea of the appearance of a house which is being designed for you by an architect, the sensible thing to ask for is a sketch, a picture in two dimensions of what is really three-dimen-

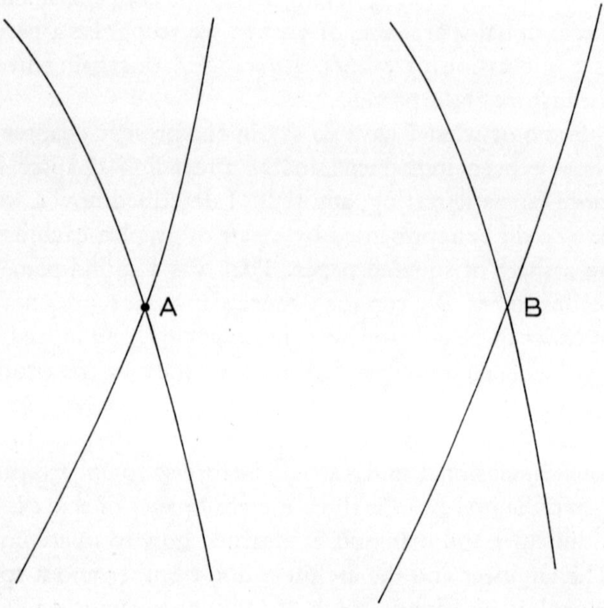

Fig. 7.1

Intersection (A) and non-intersection (B)

sional. The sketch is, in fact, a projection on an arbitrarily chosen plane.

That is a good idea when it comes to space-time – make a projection on an arbitrarily chosen plane. An event then appears as a point on the paper and the world line of a particle appears as a curve. Such a *space-time* diagram is a great help to thought, provided it is treated with some caution. In particular, you have to remember that a dot on the paper does not represent a single event, but a lot of events, all projected into that single dot, and so two world lines may appear to intersect without there being an actual intersection. However, it is easy to take care of this by using a blob to indicate an actual interesection of world lines, omitting a blob when there is only an apparent intersection, as shown in fig. 7.1.

It is amusing at this stage to take a piece of paper, draw some curves on it, and ask whether the diagram represents something possible or something absurd. Consider fig. 7.2. It shows a world line intersecting itself. That means that the *same* event occurred twice in the history of a particle as represented by the curve. Is that possible?

It is not an easy question to answer or even to understand. A man may complain that his wife gives him the same dinner every day of the week, but what he means is that all the dinners are very like each other, not that it is the *same dinner* (Heaven help us!) eaten seven times over. I am inclined to rule out fig. 7.2 as a physical impossibility, but only for reasons involving concepts not introduced so far. There is certainly no contradiction in the concept of a closed curve drawn in four-dimensional space-time. The fact is that not all curves can be interpreted as possible world lines of particles.

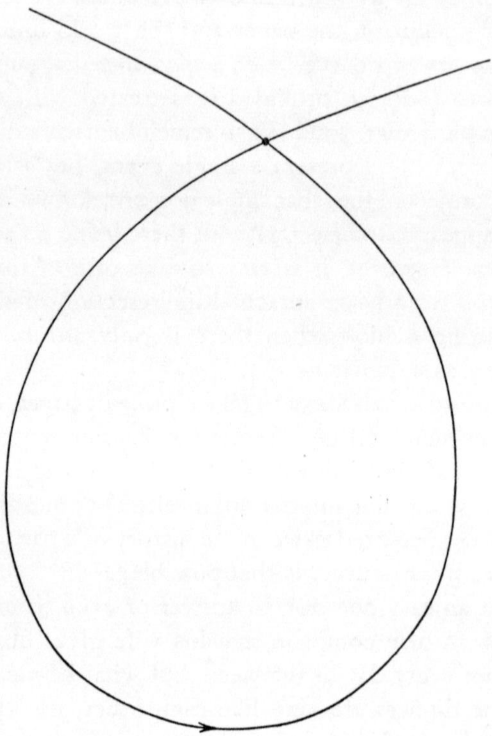

Fig. 7.2
Absurd or not absurd?

Now for the arrow of time, already introduced in the preceding chapter. This is a matter of before-and-after, and I think it is best to deal with it first in the R-world and then refine actual experience into a concept suitable for the M_2-world of relativity.

Many people keep diaries in which they record events which have happened. Suppose, to make things simple, that the diary allows a whole page per day. As I write this, the date is 3 November 1969, and that is (for the time being at any rate) a red-letter day. All preceding dates are in the past, and on the pages allotted to them there are entries of events which I experienced, directly or indirectly. All subsequent dates are in the future and the pages are blank. The diary is a sort of model of my world line, with an arrow of time pointing from the filled pages to the unfilled pages – from the past to the future. In between stands the present – 3 November 1969.

I can do better than that. According to my watch, the time is 2.27 p.m., and so I can push the earlier part of this day into the past and the later part into the future. I have reduced the present from one day to one minute. If I look at the seconds hand of my watch, I can do better still, and so one comes to think of the present as an instant with no duration at all.

Once I do that, I pass from the R-world into an M-world, the R-world knowing no instants any more than it knows Euclidean points on a sheet of paper – only smudges and scratches. For I am a brain inhabiting a body of some considerable extension and might have serious difficulty in deciding the time-order of two needle-pricks, one prick being on my left foot and the other on my right hand. I would be tempted to seek the advice of a physiolo-

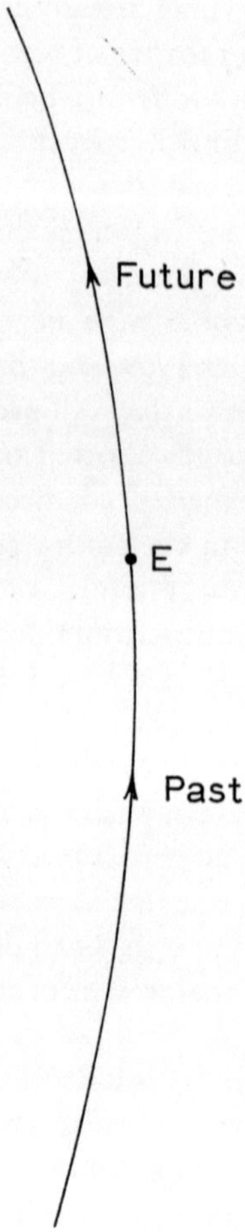

Fig. 7.3

The arrow of time on the world line of a particle

gist about the speed of signals along nerves, but that would introduce complications into what should be a very fundamental matter – for what is *speed* but *distance* divided by *time*, and we are groping towards an understanding of time?

The idea that events in the history of an individual can be put into a time-order (past – present – future) is not of course peculiar to relativity. But there is a solemn warning not to rush ahead too far and say that *any* two events can be put into a time-order, as is in fact done by following Newton into the concept of absolute time. More cautiously, we say that there is an acceptable time-order for events *in the history of an individual*. In the R-world there is a bit of fuzziness about this time-order (as with the needle-pricks), but our experience in the R-world gives us a guide to help us in setting up a suitable concept in the M_2-world of relativity.

Think then of the world line of a particle. It consists of all events in the history of the particle, and for the present we think of no events other than those. The concept of *time-order* amounts to recognising that, if A and B are two events on the world line, then either A is *before* B or B is *before* A; in the former case, we say (using a different word) that B is *after* A, and in the latter case A is *after* B. Any event E on the world line divides it into two parts, the *future* consisting of all events after E and the *past* consisting of all events before E. The *present* is reduced to the event E itself. The *arrow of time* points from the past into the future, as indicated in fig. 7.3.

The above statement merely states rather formally what most people would accept without much question. The important thing to note is that nothing is said about the time-order of two quite arbitrary events – they must both be on the world line of a particle if the words *before* and *after* are to be used in reference to them. Later on, I shall make those words more elastic, but the time is not yet ripe.

The arrow of time is linked up with two other matters, increase in entropy and causation. The former refers to the fact that nature seems to prefer chaos to order (it is easier to break a cup than to mend it when broken), and the arrow of time is usually (but not always) found to point in the direction of increasing chaos. But, important as this is, I don't feel that this is the place to discuss it further as it is not an essentially relativistic idea, and so I pass on to causation, a word about which we all have general commonsense thoughts and these can be refined into concepts for our M_2-world.

You must forgive me if I keep jumping from R-world to M-world and back again, but that is the way physical thought proceeds. I have no sooner refined a particle into a moving point than I want to drag that concept back into the R-world. There are many situations in physics where the size of a body doesn't seem to matter much. In discussing the motions of the planets, it is customary to treat them as if they had no size – as if they were particles. In plotting the course of a ship or aircraft across oceans, its size doesn't matter (docking or landing is of course a different matter). We know well that lots of things are going on in a ship or aircraft, and that what we are treating as particles actually have a rich internal structure, but that doesn't matter in the present connection. One particle-ship can send a signal to another particle-ship, and as a result the course of the latter is altered. The signal has *caused* the alteration in course. The man who ordered the signal to be sent, *caused* it.

We freeze the pattern of events into a four-dimensional space-time, but that does not mean that causation is ruled out. It is legitimate to imagine that a particle in the M_2-world sends out signals and that these are received by another particle, which may send a reply. It is legitimate to become a little anthropomorphic and think of a particle as alive. In fact, the particle becomes a tiny *observer*, but an observer (this is most important) who observes only events on his own world line. If he is to say anything about another particle, it must either collide with him or send a signal.

Signals are so important in relativity that the subject might well be called the 'theory of signalling' – it would be less confusing than the accepted title. The most important signals are light-signals, or more generally electromagnetic signals, and I shall concentrate on them. Now please do not ask how *fast* such signals travel. If you do, I shall counter by asking you what *fast* (or *slow*) means, and you will probably (but I hope not) spill out a lot of Newtonian stuff about *distance* and *time*. In that case there is nothing for me to say but that we are not in contact with each other – you are still in the M_1-world of Newton while I am in the M_2-world of relativity and I regard your words as (at least for the present) meaningless in my context.

Light-signals have a complicated structure, involving Maxwell's partial differential equations, but for our simplified M_2-world we may think of a light-signal as carried by a thing called a *photon*. This thing has a world line and we may put an arrow of time on it, pointing from its emission towards its reception.

Fig. 7.4 shows the world lines of two particles (C and D). At the event P, C sends out a signal which strikes D at Q, and is promptly returned, striking C at the event R.

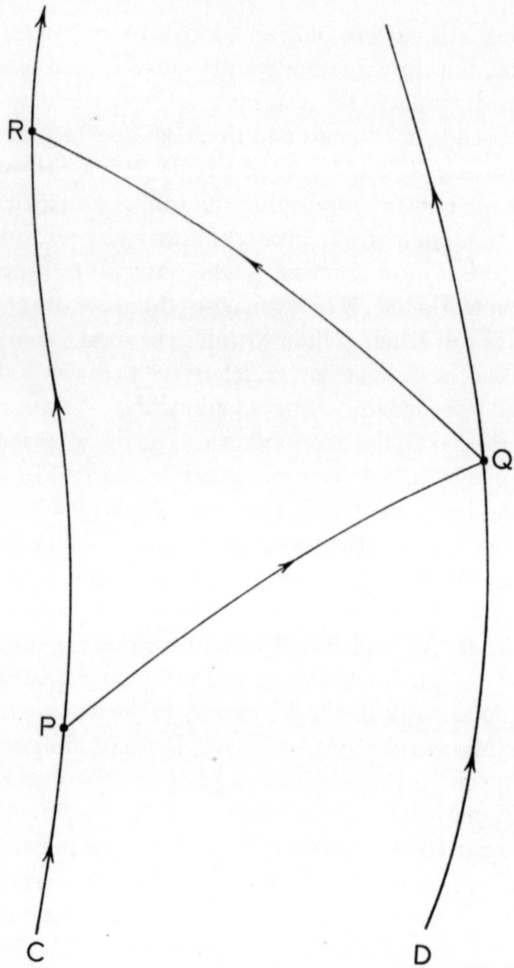

Fig. 7.4

A signal is exchanged

Consider this diagram from the standpoint of causation. The exchange started with the event P. The observer C might very well have decided not to send a signal at all. But he did, and the event P caused a signal to be received at Q. Then D might have decided not to send a signal back. But he did send it and so caused the event R. Thus, in fact, P caused R, and the diagram is sensible in that cause came *before* effect.

I am in the habit of drawing the arrow of time upward on the page, but it is merely a habit – there is nothing mandatory about it. As it stands, fig. 7.4 is an excellent common-sense diagram, and the captains of two ships at sea would recognise it as such as soon as they realised what the diagram was intended to represent. But what happens if you reverse an arrow or two?

Try reversing the arrow on QR. Now we see C sending two signals to D, both arriving at the event Q. For reasons which will appear later, this could not be done with light-signals in vacuo, but it could be done, possibly, if one signal were a sound-signal. There is nothing really absurd about the modified diagram.

But now reverse the arrows on both QR and PQ. We see C sending out a signal at R; this is received by D at Q, and returned, reaching C at P. Since we did not reverse the arrow on C, this is indeed a surprising situation. For now R is the cause of P, and the cause came *after* the effect. That will not do at all. The modified diagram must in this case be firmly rejected. The test for rejection of this and like diagrams may be stated thus: *there must be no closed loops, when you follow the arrows*. We must classify as impossible the diagrams in fig. 7.5, and likewise fig. 7.2 must be rejected as absurd.

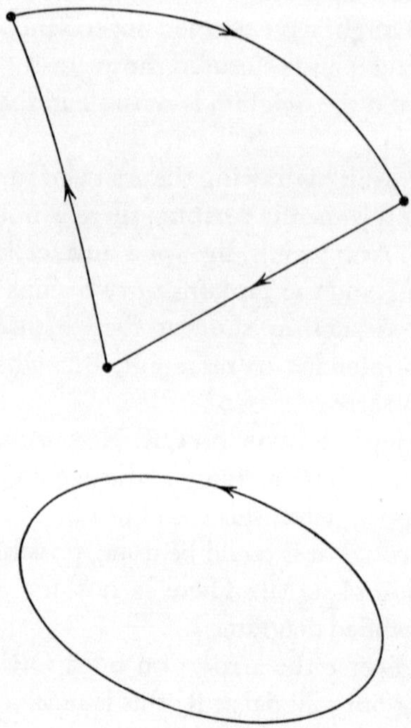

Fig. 7.5
Impossibilities

What would Newton say to all this? First, he could not object to four-dimensional space-time, although he would prefer to see it sliced into three-dimensional space and one-dimensional time. Secondly, he could not object to the world lines of the two particles in fig. 7.4. Thirdly, he would have to admit that an observer could know nothing about events off his world line except through the medium of signals. Fourthly, he would be ready to accept fig. 7.4 as a representation of light-signals exchanged. But having admitted all that, he would say we were foolish to use light-signals. It would be better to shake a massive fist, and this would cause an instantaneous gravitational effect, which could be returned immediately (another massive fist shaken). With this instantaneous signal, fig. 7.4 would collapse into fig 7.6, in which P and R are now one single event.

If Newton said all this, the correct and polite reply would be to insist that his view was perfectly correct as far as his M_1-world was concerned, for this incorporates absolute time or, equivalently, instantaneous signalling. But the M_2-world of relativity rejects the concept of instantaneous signalling and insists that the best signal is the light-signal in vacuo. There is no reason to quarrel over this, for everyone is free to make his own M-world. The crunch comes when we compare M-worlds with the R-world to find out whether there is a good agreement. On the basis of a good deal of accumulated evidence, mostly indirect, it is held that there is no signal known to man which does what Newton's instantaneous signal does in fig. 7.6, in fact no signal better than a light-signal. If such a signal were discovered, relativists would have to do some hard rethinking to fit the M_2-world to the R-world.

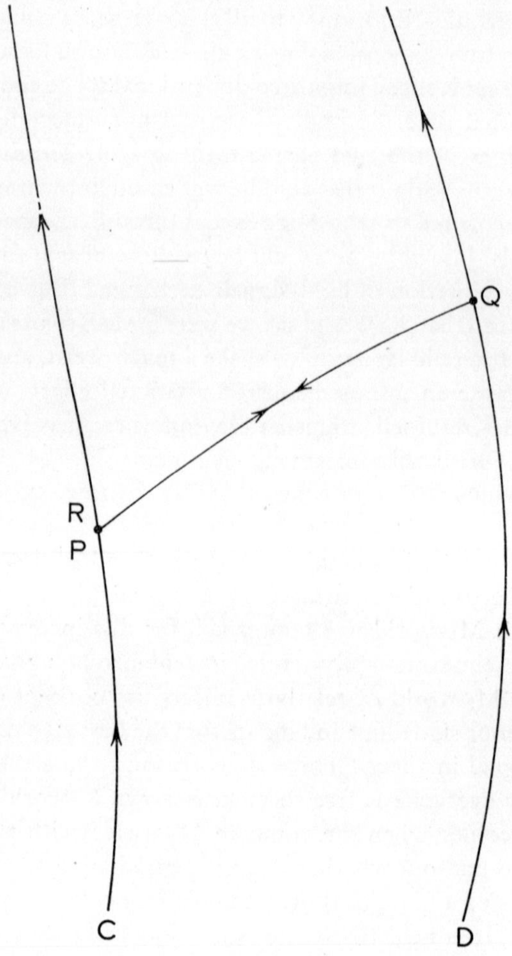

Fig. 7.6
Newton's instantaneous signal

Vulgarly you may say (but I rather hope you won't) that no signal can travel faster than light. Such a statement merely makes easy things hard. It is the round-trip that matters, not the journeys out and back separately, and you can safely postpone for the present any thought about speed or velocity, which are indeed pretty hard things to think about.

This is the moment to delete a word from your dictionary – the word *simultaneous* – or, if you do retain it, it will be in a very emasculated form. In the Newtonian picture (fig. 7.6) the events P and Q are simultaneous, and the word is perfectly good (indeed fundamental) in the Newtonian context. But it would be wrong to say, relative to fig. 7.4, that there is any simultaneity of events. For causal reasons, we would say that R is after Q and Q after P, and that might lead us to think that there must be some event on C, between P and R, which might be fitly said to be simultaneous with Q. It is possible to rig up a definition of sorts (using the concept of relative time, to which we shall come in chapter 9), but it is a poor thing, not to be taken too seriously in a general context. So please forget the word *simultaneous*, except perhaps in the trivial sense that two events are simultaneous if they coincide.

With the rejection of simultaneity in any absolute sense (except this trivial one) we have indeed come to the parting of the ways, Newton one way, Einstein the other.

CHAPTER 8

The Mary Jane and the Penelope

Two ships, the Mary Jane and the Penelope, left the port of Azzam together, both bound for Bezam. They separated but kept in radio contact.

One morning, as the skipper of the Mary Jane was sitting down to his breakfast, he wondered whether he was up before the skipper of the Penelope, so he called his wireless operator and told him to send a signal to the Penelope to ask in courteous terms whether her skipper had yet sat down to his breakfast. All this happened in the twinkling of an eye and the message was despatched from the Mary Jane at the very instant when her skipper took up his knife to decapitate his egg.

The wireless operator on the Penelope wasted no time. He rushed to the cabin where the skipper took his breakfast, saw he was not there yet, rushed back to his transmitter and sent off a message to the Mary Jane to say that the skipper of the Penelope had not yet sat down to breakfast.

The news delighted the skipper of the Mary Jane. 'I breakfasted before that other fellow,' said he, addressing a portrait of Einstein screwed to the wall of his cabin; the portrait seemed to nod approval.

Next morning the operation was repeated, but with a different result, for back from the Penelope came the message: 'Skipper already started and halfway through his egg.'

'Damn!' cried the skipper of the Mary Jane. 'That other fellow beat me to it today!' But looking at the portrait of Einstein, he detected no approving nod, only a slight grin. What was wrong? It was clear that the skipper of the Penelope had started his breakfast before the signal from the Mary Jane arrived. But then light dawned on the skipper of the Mary Jane.

'The signal took time to travel,' he said to himself, 'and perhaps I did start my breakfast first after all. I started my breakfast just when the signal went out, and that was certainly before it arrived at the Penelope. He, too, had started his breakfast before that. But that does not put one breakfast before or after the other.'

He realised that he had quite a problem on his hands, the problem of simultaneity. It was no part of his duty as skipper to bother about such things, but he could not give it up and it kept him twisting and turning in his bunk all night and every night. It seemed to him of paramount importance that he should find out what precisely the skipper of the Penelope was doing at the very instant when he sat down to his breakfast in the Mary Jane. He realised that there would be no problem if they were both on the same ship, sitting down side by side at the same table. The trouble arose from the fact that a signal took time to travel, and he searched the ship's library to find out whether there was any better signal than wireless, but all the books said there was none. A light-signal would travel at the same

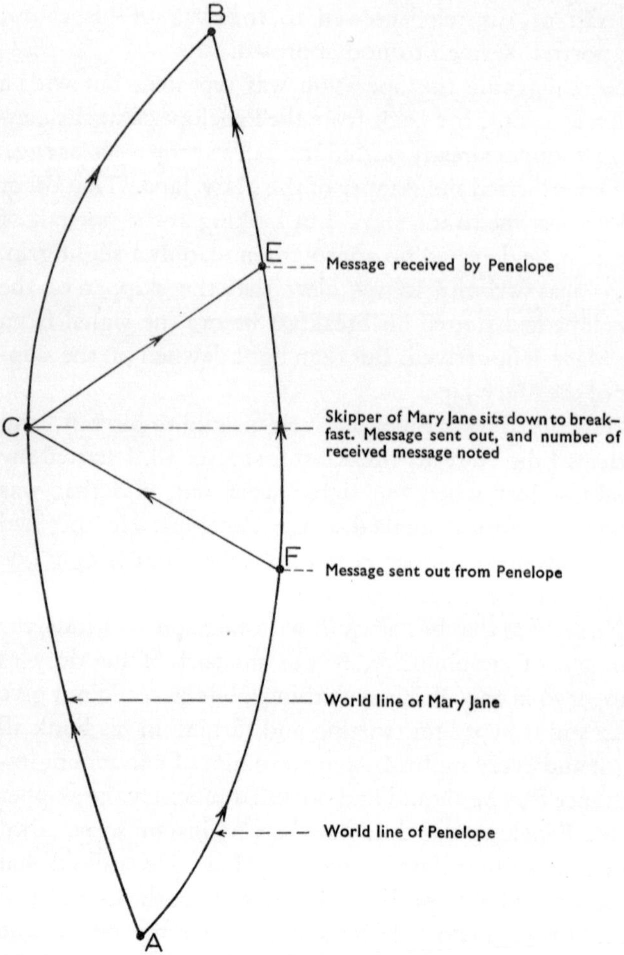

Fig. 8.1

The Mary Jane and the Penelope

speed and a sound-signal would be slower. There was no help there.

He decided to enlist the cooperation of the skipper of the Penelope and he sent long messages explaining what the problem was. The skipper of the Penelope thought that the skipper of the Mary Jane had gone off his head but was willing to humour him. So it was arranged that the Penelope would send a succession of signals to the Mary Jane, each signal having a number (1, 2, 3,...) coded into it, so that the departure of the signals from the Penelope could be identified and also their arrival at the Mary Jane. Thus the instants of emission of signals would stand in the time-order 1, 2, 3,... and likewise their arrivals would stand in the same time-order.

But would they? It had to be tested, so the Penelope sent a signal 1, followed by a signal 2, and they were received by the Mary Jane in that time-order. It would have been very confusing if the signals had played hare-and-tortoise, but that did not happen. The signal sent off second never arrived first.

With that settled, they got down to business, but as this is getting rather complicated, I offer a space-time diagram (fig. 8.1) to make things easier to follow. This is a space-time diagram, not a chart of the sea. The two curves joining A and B are the *histories* of the two ships (their world lines), A being the event of their departure from Azzam and B the event of their arrival at Bezam. A signal sent from the Penelope to the Mary Jane appears as a line sloping up to the left (like FC) and a signal going the other way slopes up to the right (like CE).

C is the event of prime interest – the skipper of the Mary Jane sits down to his breakfast. A succession of signals is being sent from the Penelope and one happens to arrive at the event C; it was sent out at the event F. From C a signal is immediately sent back to the Penelope and arrives at the event E.

Now the two skippers had agreed that neither would steal a march on the other – they would sit down to their breakfasts simultaneously, or at the same time, or whatever you like to call it. It was further decided that the skipper of the Mary Jane would take the initiative and sit down when it pleased him, while the skipper of the Penelope would adjust his breakfast suitably in order to secure simultaneity. So the messages passed to and fro in the manner aforesaid, but it was not until the event E that the skipper of the Penelope was able to assuage his hunger, for it was only then that he received the news that the skipper of the Mary Jane had started his breakfast. And the event E was obviously too late.

So they had to hit on another plan, and it was decided that the skipper of the Mary Jane would breakfast at his pleasure as before, while the whole crew of the Penelope would breakfast in quick succession, one after another. Surely, it was felt, some member of the crew of the Penelope, if not her skipper, would happen to breakfast simultaneously with the skipper of the Mary Jane and accordingly a sweepstake was arranged. But the plan fizzled out, because all that could be said was this: any sailor who breakfasted before the event F was too early and any sailor who breakfasted after the event E was too late. As for those who breakfasted between F and E, none could be picked out as winner of the sweepstake, and so the prize was divided equally among them. One sailor protested that he should receive the whole prize because he breakfasted halfway between F and E, but, since he could not say what he meant by 'halfway', he was deprived of his share in the prize and put in irons for insubordination.

When all this was reported to the skipper of the Mary Jane, he went into his cabin and knelt down before the portrait of Einstein.

When they got to Bezam the two skippers remembered that they had forgotten about the perfectly good chronometers they had on board, and wondered whether they should have used them to solve the problem of simultaneity. But of that more later, on p. 111.

CHAPTER 9

Time

When maids got up at seven,
And gentlefolk at ten,
The real aristocracy
Just took their little cup of tea
And settled down again.

Newtonian time was universal; relativistic time is, at it were, personal. My time is not your time nor your time my time. At least not unless we keep close company or decide to neglect sufficiently small differences.

I decided to head this chapter with the single word TIME. I might have said PROPER TIME, this being the usual translation into English of the German EIGENZEIT, but I refrained for what I consider sufficient reasons. One reason is that *proper* is a poor translation of *eigen* – we do not speak of a man's *proper head* when we mean his *own head*. The second reason is that if I say 'proper time', you will quite properly ask what *time* is, the naked word without qualifying adjective. The fact is that this *proper time* IS *time* in the deepest sense, and if you want to use the word in some other, less fundamental, sense, then it is up to you to supply the suitable qualifying word. For example, you may wish to use the word *time* for one of the four coordinates in space-time; but, if so, you should say *coordinate-time*, since this is only a second-rate sort of time.

But here I am running on about *words* when what you want is a *concept*, the concept of time. It is true that I have spoken of the time-order of events – one event in the history of a particle is before (or after) another – but a concept comes to life only when you say how it is to be measured, and anyway *time-order* is not *time*. The concept of time is not a thing to be rushed at, so let us talk about it first in familiar terms.

The ordinary man or woman keeps time by means of a watch or clock, correcting it occasionally by time-signals given on radio or TV. Those time-signals are controlled by astronomical observation of the apparent rotation of the heavens, which apparent rotation is due to the real rotation of the earth. It appears then that the measurement of time is wrapped up with celestial mechanics, and it would be correct to say that time, as ordinarily understood, is the independent variable t which occurs in the Newtonian equations of motion which govern celestial mechanics and that its measurement involves the solution of those equations. In the simplest application of this idea, consider a body moving under no forces: then time is proportional to distance traversed. In the Newtonian scheme of things, *time* is thus to be measured in terms of *distance*, and distance is a primitive concept – you measure it with a rigid measuring rod.

If you glance back at the preceding paragraph, you will notice that I started off in the R-world, with people looking at watches and clocks, noting time-signals, and observing the rotation of the heavens. But then I slipped into the M_1-world of Newton, with equations of motion, and independent variable t, and distances measured with a rigid rod. It is very hard to keep those two worlds apart, but we must make the effort because the M_1-world of Newton is a sticky fly-paper for anyone who aspires to fly into the M_2-world of relativity.

So let us wipe our feet and wings and retreat into the security of the R-world in which things really happen, and from which the mad game of mathematics is excluded, at least for the present. Does it not seem strange that it should be necessary to bring in the apparent motion of the stars in order to deal with such a fundamental concept as that of time? Would not a mechanical clock do the trick, provided it was very well made? The essential criterion seems to be that the mechanism should do the same thing over and over again under precisely the same conditions. The primitive water-clock and the hour-glass are crude examples. But they are too crude. It is better to use a clock mechanism with balance wheel and escapement. But once again we seem to lose sight of a simple concept in a complexity of mechanism.

It is a relief to know that we can fall back on atomic clocks. We may find it hard to build up a realistic picture of their mechanism, but that does not matter. In broad outline we say that an atom may be excited to certain energy-levels, and that when it falls from one level to a lower one, it emits radiation with a frequency (number of vibrations per second) determined by the type of atom used and the two energy-levels involved. We may say that an atomic clock ticks with a definite frequency – it keeps time – and this agrees with remarkable accuracy with the usual astronomical time. Indeed it is becoming the vogue to *define* the unit of time in terms of an atomic clock, and to regard astronomical time as secondary.

Although physicists slip into a mathematical M-world to study the theory of an atomic clock, the clock itself is as much in the R-world as your wrist-watch is. As usual, it is natural to take something from the R-world and refine it into a concept in that M-world with which we are concerned. So, into the M_2-world of relativity, we take over the concept of an atomic clock as the measurer of time, and this we do by imagining that any particle may be provided with one. Since these are the only clocks in our M_2-world, I shall drop the adjective *atomic* and call them *clocks*, without qualification. Some people like to say that a *standard clock* measures *proper time*, but I shall say more simply a *clock* measures *time*, meaning precisely the same thing.

There has been much vague talk about time through the ages, and, if I have succeeded in getting my concept of time across to you in a couple of pages, then I am entitled to feel agreeably elated. I say *my* concept, not out of vanity, but out of modesty. I cannot be sure that Einstein would have approved of what I have said, and there may be a number of living relativists who would look at time differently. Yet, on the whole, I think I am being orthodox. Certainly I am not being deliberately heterodox for the sake of being different. I have given you my concept of time in the M_2-world of relativity, and, if anyone has a different concept, let him come forward and explain it. Whether I shall understand his concept or not is quite another matter.

We need a space-time diagram to consolidate the concept of time. Fig. 9.1 is a dual-purpose picture: it illustrates the

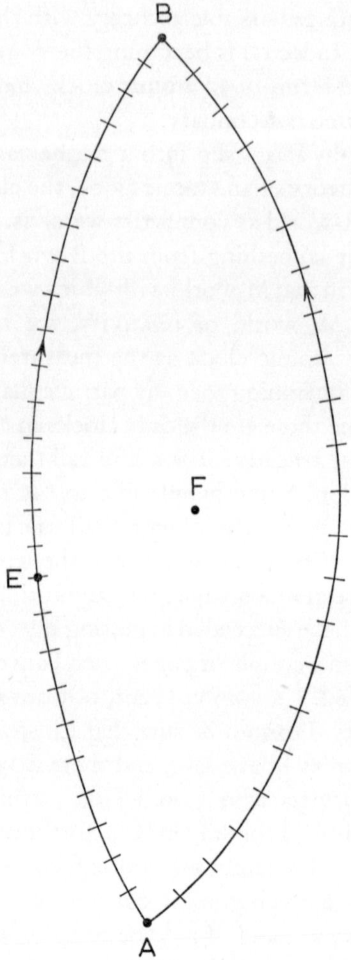

Fig. 9.1
The concept of time illustrated and the 'clock paradox' exploded

concept of time and it serves to explode an absurdity which goes under the name of the 'clock paradox'. To illustrate the concept of time, it would have been enough to draw one world line. Concentrate attention then on the world line to the left. It runs from event A to event B and carries the arrow of time, indicating time-order. The marks on the curve represent certain events – the ticks of the clock carried by the particle. For simplicity I consider A and B to be tick-events, and (to be able to talk simply) I have put in 25 intermediate ticks.

As we advance, in the sense of the arrow, from one tick to the next, time has increased by one unit. We may choose the zero of time anywhere on the world line. Choose it at A. Then the time of the event B, according to the clock on the left-hand world line, is 26 units. Likewise the time of the event E is 10 units. That should illustrate sufficiently the concept of time as far as the left-hand world line is concerned.

Now take the right-hand world line, on which I have marked 21 intermediate ticks. If we take A again to be at zero time, the time of the event B is 22 according to the clock on that world line.

Have I made a mistake? Should I have inserted the same number of ticks on each of the two world lines? No, certainly not. True, it would make the time-interval between events A and B independent of the history of the clock that recorded the two intervals, and that would appear a nice simple axiom to put into our M-world. But relativistic time is personal, *not* universal, and if we insisted on making a time-interval independent of the clock-history, we would in effect be insisting on absolute time, and we would find ourselves back in Newton's M_1-world and not in the M_2-

world of relativity. Thus, if we draw a space-time diagram like fig. 9.1, marking in clock-ticks, we should in general put *different* numbers of ticks on the two world lines. (I say 'in general' because there may be symmetrical conditions by virtue of which the two numbers are the same.)

Some people have found something repugnant to reason (or to their concept of what relativity is) in making time-intervals depend on clock-histories, and they call this the 'clock paradox'. It is not a paradox at all, but a fundamental part of relativity.

In addition to the two world lines, I have inserted in fig. 9.1 an additional event F, not on either world line. What is the time of F? I ask this silly question merely for the purpose of pointing out its silliness. Time is personal, not universal, and F has no personality in the sense that there is no indication of a clock whose history includes the event F. It may be possible to have a clock contain both A and F in its history, or it may not (there is an important restriction to be discussed later). But then the time of F (with zero time at A) will depend on the world line chosen to connect it with A, and so the question, as asked, has not got any significant answer.

I have been talking about particles carrying clocks, and it may seem absurd to think of a moving point (of no size) carrying a time-recording mechanism. But I think that any absurdity there may be arises from trying to mix the M_2-world with the R-world. From the standpoint of the M_2-world we have a world line or curve in four-dimensional space-time with tick-marks on it, and there is nothing absurd about that – it is a perfectly valid mathematical concept. From the standpoint of the R-world, we see perhaps a ship in the ocean (the ship being very small in comparison to its distance from another ship) and on the ship a very accurate chronometer – no absurdity there either. Or, for some astronomical purposes, the 'particle' might be the earth and the clock might be the clock in the astronomer's observatory.

But if there is actually no absurdity in talking about a particle carrying a clock, the same cannot be said for a photon (or light-signal) carrying a clock. Nevertheless I would like to talk about it, regarding it as a limit unattained, even in the M_2-world. Perhaps it would be better to wait until we get to the mathematical structure of relativity, but the salient facts (best regarded as mathematical deductions from that structure) are so interesting that I have good excuse for proceeding with them now.

To get the feel of the thing, start in the R-world with a powerful bomb. The bomb explodes, and an observer sees the flash and manages to receive the fragments of the casing without injury to himself. He thus experiences a sequences of events – seeing the flash and receiving the fragments. The flash arrives first.

Now idealise that picture. Draw the world line of the observer, as shown in fig. 9.2, in which I have not bothered to mark in the ticks of his clock. Now mark the event of explosion E, and the world lines of the fragments and of the light-flash. All this is pretty near to reality, but now I am going to suppose that each fragment carries its own private clock, set to read zero at the event E, and ticking away as it flies. Further (and this is indeed fantastic) suppose the bomb so powerful that some of the fragments reach the observer only just after the flash – 'just after' referring of course to time-measurement by the observer's clock. (We might say that some of the fragments travel

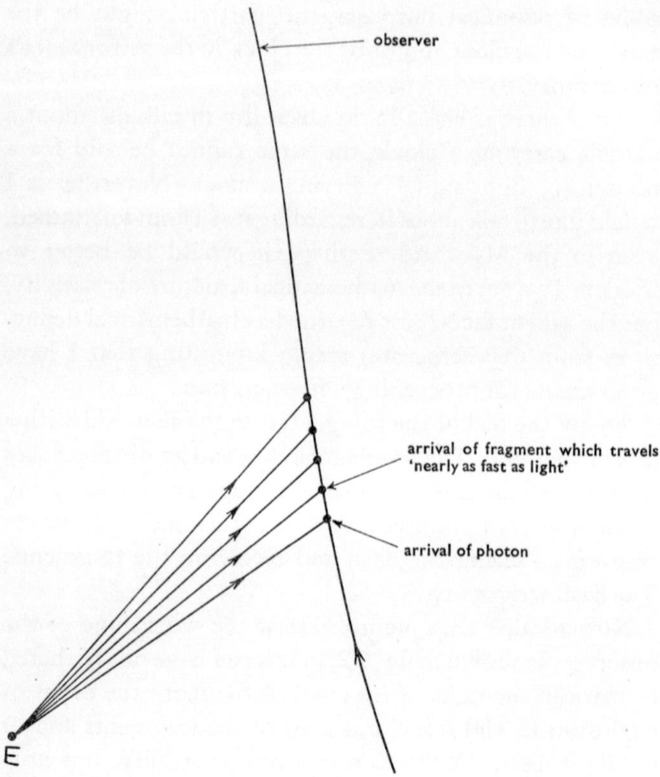

Fig. 9.2

The clock on the photon does not tick

'nearly as fast as light'.)

Now what we are concerned with is the set of readings of the fragment-clocks which arrive just after the flash arrives, and the fact is that (in the M_2-world) they read as near to zero as doesn't matter. They don't get the chance to tick at all in their flight from explosion to observer, and this we express in a legitimately simplified form by identifying the photon with these fast particles and saying that *the photon clock does not tick at all.*

I shall round this off by accompanying the Mary Jane and the Penelope on another voyage from Azzam to Bezam. This time the skippers plan to keep radio silence and use their excellent chronometers to ensure that they breakfast simultaneously. As the ships lie side by side at the dock in Azzam, the clocks are carefully compared and are found to run true with each other. As the ships are together, there is no question of sending signals.

The two ships set out and the two skippers breakfast every morning at 8 o'clock, that being the agreed time. They arrive at Bezam at night and tie up side by side at the dock. But when the skipper of the Mary Jane glances through the porthole as he is sitting down to breakfast next morning, he sees that the skipper of the Penelope is already at work on his egg. 'The cheat!' he cries and rushes across the gangplank with murder in his heart. He expects to find the Penelope's chronometer reading 8 hours, but to his amazement it reads 8.10.

'Throw that bloody old chronometer overboard!' he shouts at the skipper of the Penelope. 'It's no damn good!'

But the skipper of the Penelope has been taking a special course in nautical relativity, and he knows what has happened. The two ships started together and arrived together,

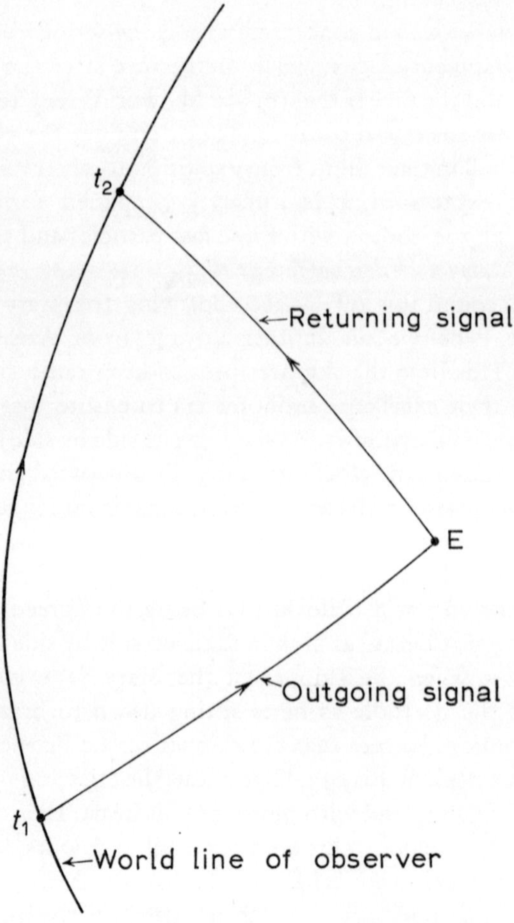

Fig. 9.3

Relative distance and relative time

and had indeed travelled on the same course. But whereas the Penelope had ploughed steadily forward, the Mary Jane had trouble in the engine-room and was proceeding in a stop-go manner. The Mary Jane was certainly not a photon, but she had been behaving in a rather photonic manner with those sudden bursts of speed to make up for her pauses, and the effect of this was to make her chronometer tick less often than the Penelope's.

I need hardly say that Bezam is a long way from Azzam.

* * *

Fig. 9.3 shows the world line of an observer. He sends out a light-signal at time t_1, as recorded by his clock. E is an event on the world line of that signal. The signal is reflected at E and returns to the observer at time t_2, as recorded by his clock.

Only two numbers, t_1 and t_2, are involved, and in terms of them the observer may, if he so chooses, define distance and time for the event E. But, since his world line is intimately tied up with the observations, it is desirable to indicate this by inserting the word *relative*, and accordingly we write

$$D = \text{relative distance} = \tfrac{1}{2}(t_2 - t_1),$$

$$T = \text{relative time} = \tfrac{1}{2}(t_2 + t_1).$$

We could hardly think of anything simpler. The relative distance is half the time-interval between the departure of the signal and its return, and the relative time is the arithmetic mean of the two recorded times.

The situation depicted in fig. 9.3 is by no means fantastic. It occurs when you shave in front of a mirror: the world line is that of your eye and E is an event in the history of the mirror. The relative distance is half the time taken by light to travel from your eye to the mirror and back again. We are slow-moving creatures (you would not have time to wink before the signal returned to your eye) and so we find it convenient to use different units for distance and time and write

$$D = \tfrac{1}{2}c(t_2 - t_1)$$

where c is a very large number if we decide to measure D in centimetres and time in seconds, viz. $c = 3 \times 10^{10}$. I deliberately refrain from saying that this number c is the 'speed of light' since that would only cause confusion. I prefer to retain the simpler formula; in other words, I prefer to define the unit of distance directly in terms of the time taken by light to go out and back – I repeat *out and back*, not *out* only or *back* only, for we have set up no means of measuring such time-intervals. Indeed since 'a photon clock does not tick', we would find ourselves in the peculiar position of saying that the outgoing signal took no time at all, and the returning signal took no time at all, but the total trip did take a certain time, viz. $t_2 - t_1$. If you find that confusing, it is because your Newtonian intuition is still active.

It may be as well to wrestle a bit with that Newtonian intuition in which the concepts of distance and time are built-in. Thus, if a Newtonian physicist is thinking about the transmission of a signal to the moon and its return, he *starts* with the concept that at a certain *instant* the moon is at a certain *distance*, and any observations he may make are aimed at measuring *things that exist*. From the relativistic standpoint, those things just don't exist. There is no concept in relativity of the distance from the earth to the moon – it is a matter of definition, and we *define* it in the simplest way by writing $D=\frac{1}{2}(t_2-t_1)$. Nor is there any concept in relativity of the instant (in the observer's time-scale) when the signal strikes the moon, and so we *define* it by $T=\frac{1}{2}(t_2+t_1)$. There is nothing sacrosanct about those definitions of D and T. You may think up others, but these are the simplest possible and I shall stick to them.

CHAPTER 10

The line-element and curvature of a surface

It is very hard to escape from Newtonian ideas, for they permeate ordinary life and by far the greater part of scientific education. If it were possible to analyse my own brain, I believe that the Newtonian part would outweigh the relativistic part, and I have to confess that, on the whole, I prefer Newtonian problems to relativistic ones, partly because the mathematics is easier (and hence there is more chance of an interesting solution) and partly because the flow of intuition between the R-world and the M_1-world of Newton is freer than between the R-world and the M_2-world of relativity. If someone threatened to shoot me, I would take intuitive evasive action and, in so far as that action was based on any theory, it would surely be Newtonian theory. This might trouble my conscience in that I believe relativistic theory to be a better approximation to reality, but in an emergency there would be no time to deal with its harder mathematics, and at any rate it would make only an insignificant difference in the prescribed course of action unless my assailant had, relative to me, a speed comparable to that of light, an unlikely contingency unless it was a case of cosmic rays as bullets.

Let us then frankly admit that thinking about relativity puts one in a state of mental tension. It is so easy for some Newtonian idea to leak out from the subconscious and vitiate an otherwise (relativistically) impeccable discussion. Only under two conditions do I find myself secure against that leak: first, when I make a space-time diagram (they are unusual in Newtonian discussions) and secondly when immersed in a mathematical argument in which the various symbols are natural to relativity. In the domain of mere words, I feel insecure and liable to slip into Newtonian ways of thinking unless I exercise great care. The danger is greatest when I wander off into the realms of fantasy.

I say all this to prepare you for a bit of good news: in the present chapter we can relax, for the subject is the ordinary geometry on which we were reared. It is not even necessary to draw a sharp distinction between R-geometry (model surfaces made out of leather or plaster, measurements on such surfaces made with pieces of string, and so on) and M-geometry (lines and surfaces of no thickness, all procedures strictly logical, and so on). We can, as I say, relax; for this chapter, although in one way essential to the relativistic purpose, exists only to provide a basis for analogy.

Think, then, of a sculptor chipping away at a block of marble and ending up with a statue, a solid body with a curved surface. How is that surface to be accurately described?

The most systematic plan is to take axes of coordinates x, y, z and describe the surface by means of an equation $f(x, y, z) = 0$. If the surface is a sphere of unit radius, the equation reads

$$x^2 + y^2 + z^2 - 1 = 0,$$

and if it is an ellipsoid with semi-axes 3, 3, 2, the equation is

$$4x^2+4y^2+9z^2-36=0 .$$

No such simple equation would describe the surface of a statue, except perhaps a very modern one, but it can hardly be denied that some equation exists.

In describing a surface in this way, we regard it as immersed in three-dimensional space (we use all three coordinates in the description), but we must do better than that if we are to construct a stepping stone to relativity. It is in fact necessary to think of the surface as a thing in itself, and I cannot introduce that idea better than by referring to the case of the rich man who decided to put a statue of his wife on the lawn in front of his palatial home. He therefore engaged the services of a famous sculptor and did not give the matter another thought until one day he returned home to find the statue standing on its pedestal on the lawn. It was an excellent statue, but it had nothing on. This shocked the rich man and he suggested that at least a dressing-gown should be provided, but his wife and the sculptor both objected strongly and a compromise was finally reached – the statue should be provided with a skin-tight garment to cover it completely and accurately.

So a scholarly and cultured tailor was consulted. He said it was not as easy as might appear, that mathematicians had discussed the 'tailor's problem' and found it difficult. That problem, he explained, was the problem of applying flat cloth to a curved surface, the cloth being made of threads which were inextensible to all intents and purposes. Wrinkles were inevitable. To secure a perfect fit, it would be necessary to use a material which stretched freely in all directions, and it was not in his province to work with materials of that sort. Why not give the statue a coat of paint?

But this plan was unacceptable, and the problem was put in the hands of a geometer who said it was all a matter of the lady's line-element. He did not offer to construct the skin-tight garment, but merely to explain what was at stake.

However I do not want to get the intrinsic geometry of surfaces tied up too much with the clothing of statues, and it may be wise to refer also to cloth-covered tennis balls and the leather cases of footballs (association and rugby). Here too material is moulded to fit curved surfaces.

To put yourself in the proper frame of mind for the discussion of the intrinsic geometry of a surface, it is desirable that you should think of yourself as a two-dimensional being confined to the surface, but able to move about in it. But this involves certain embarrassments – a digestive canal from one end to the other would divide you into two distinct parts which would separate unless special precautions were taken – and it is perhaps best to settle with a compromise: let us be tiny three-dimensional bugs moving about on a surface.

We may be so regarded anyway when one thinks geographically, and the intrinsic geometry I have to speak about is quite familiar in the case of the earth, but deceptively simple, the earth being very nearly spherical.

Let us then suppose we are bugs on a surface and that we are equipped with measuring tapes to measure distances with, on the understanding that the tapes must be kept in the surface. In order to be able to discuss the geometry of the surface, we need to introduce coordinates, say (x_1, x_2), like latitude and longitude on the earth. It is important however to note that latitude and longitude are highly privileged coordinates (everybody uses them), but on a general surface there are no such specially simple coordinates. So, as often happens in such cases, we go to the other extreme and deliberately accept the fact that the coordinates on a surface are to be arbitrarily selected – we are to be prepared to change from (x_1, x_2) to (y_1, y_2), say, the y's being functions of the x's; for the x's determine some point on the surface and that point determines the y's.

It might seem impossible to say anything simple and interesting about measurements on a surface without saying just what the surface is. But it is possible, provided we are satisfied with general statements. Of such statements, the most fundamental is the following.

Let P and Q be adjacent points on the surface. Let the coordinates of P be (x_1, x_2) and those of Q (x_1+dx_1, x_2+dx_2) where dx_1 and dx_2 are infinitesimal. Then the distance ds between P and Q is itself infinitesimal, and (this is the interesting thing) its square is a *quadratic* expression in the two dx's. This means that

$$ds^2 = A_{11}dx_1^2 + 2A_{12}dx_1dx_2 + A_{22}dx_2^2\,, \qquad (10.1)$$

the three A's being functions of the coordinates (x_1, x_2). This is the *line-element* of the surface. The fact that the line-element is of that form is a consequence of the fact that the surface is embedded in Euclidean 3-space, for which the line-element may be written $dx^2+dy^2+dz^2$, but we can shake ourselves free from this Euclidean 3-space and go thoroughly intrinsic by simply slapping down a quadratic form as above and proceeding from there.

To illustrate the line-element, consider first the case where the surface is actually a plane. Then we can choose special coordinates for which

$$ds^2 = dx_1^2 + dx_2^2\,,$$

and so

$$A_{11}=1, \qquad A_{12}=0, \qquad A_{22}=1\ .$$

Secondly let the surface be a sphere and let x_1 be latitude and x_2 longitude. Then

$$ds^2=r^2dx_1{}^2+r^2\cos^2 x_1dx_2{}^2\ ,$$

where r is the radius of the sphere, and so

$$A_{11}=r^2, \qquad A_{12}=0, \qquad A_{22}=r^2\cos^2 x_1\ .$$

Differential geometry is a fascinating mathematical subject but this is no place to go deeply into it. There are, however, some more things to say in order to prepare for relativistic analogies.

Let us tidy up the rather long expression for the line-element as set out in equation (10.1). This can be done in at least two ways. I choose one of them and define A_{21} by $A_{21}=A_{12}$. Then although there are only three functions in (10.1), I can exhibit those functions in the form of an array or matrix

$$A=\begin{pmatrix} A_{11} & A_{12} \\ A_{21} & A_{22} \end{pmatrix} \tag{10.2}$$

the letter A now standing for the whole set of four (actually only three) A's. This is a *symmetric* 2×2 *matrix*; for ease of reference we may call it the *metric matrix*, since ds measures distance, and we may call the separate A's the *components of metric tensor* (see chapter 12).

But I don't want to burden the talk with a lot of technical terms, necessary as they are if we wish to pursue any detailed argument. I introduced the metric matrix in order to be able to write the line- element in the form

$$ds^2=dx\cdot A\cdot dx, \tag{10.3}$$

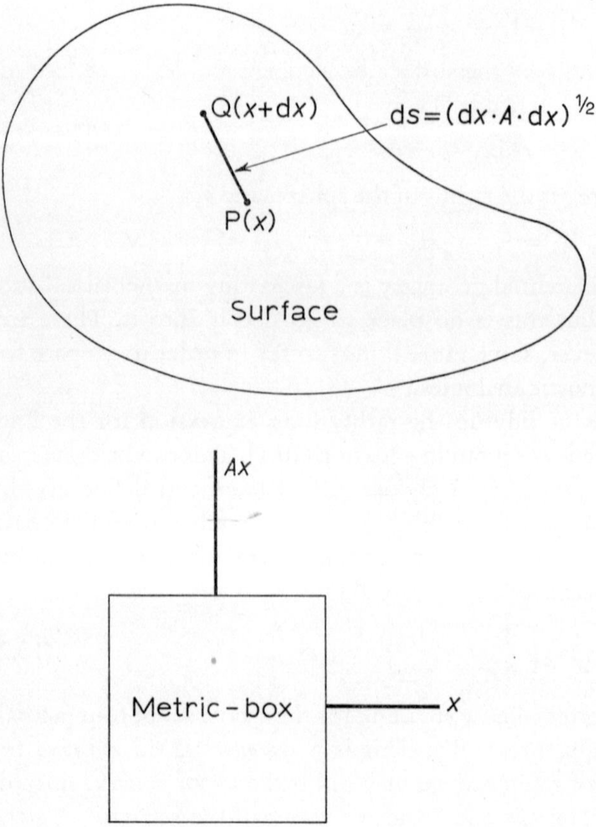

Fig. 10.1

The line-element and the metric function-box

but I don't propose to explain the structure of this formula in terms of the multiplication of matrices. Suffice it to say that (10.3) is a convenient shorthand for (10.1) – it says the same thing more briefly, which is the purpose of any good mathematical notation.

If we let the single letter x stand for the pair of coordinates (x_1, x_2), the fact that the A's are functions of the x's is taken care of by the notation Ax (cf. chapter 4), and we can sketch a function-box as in fig. 10.1 to remind us of the fact.

If you choose three functions Ax at random, you may find yourself in an absurd position. If, for example, you make the very simple choice

$$A = \begin{pmatrix} 1 & 0 \\ 0 & -1 \end{pmatrix}$$

you get the line-element

$$ds^2 = dx_1{}^2 - dx_2{}^2 \, .$$

Thus $ds = 0$ if $dx_1 = dx_2$, and, worse still, ds^2 is negative (and ds imaginary) if the second part outweighs the first. But it is absurd that the distance between two distinct points should be zero, and still more absurd that it should be imaginary. You must, in fact, discipline your choice of the functions Ax so as to make ds^2 positive, and this is expressed in technical language by saying that *the form* $dx \cdot A \cdot dx$ *must be positive-definite*. This is so obvious as hardly to need emphasis; the important thing is that it is not so in the relativistic analogue.

To get a straight line on a plane, you pull a cord tight between two points. There are no straight lines on the earth's surface. But there are great circles, and you get a great circle on a globe by pulling a string tight between two points on it. You can do this with any surface and the curve you get is a *geodesic*. Geodesics belong to the intrinsic geometry of a surface because the operation involved (tightening a string) does not require that you get out of the surface.

Any curve on a surface has a length and we write $\int ds$ to denote it, this being the mathematical way of saying that we add together the lengths of the infinitesimal portions into which it may be decomposed. Geodesics are described by the formula

$$\delta \int ds = 0\,, \tag{10.4}$$

which is merely a way of saying that if you wobble a geodesic a little the change (or variation indicated by the symbol δ) is much smaller than the wobble. In fact, the formula (10.4) says that the length has a *stationary* value, not necessarily a minimum. It is wise then to revise our definition of a geodesic and say that it is a curve of stationary length. The wisdom of this revision is apparent when we think of a sphere, for any two points on it are joined by *two* geodesics of which only one is of minimum length.

In the relativistic analogue, the minimum property disappears, but the stationary property persists.

* * *

Once upon a time there were two bugs who lived on the surface of a football, and disagreed as to whether it was an association football or a rugby football. Indeed, they disagreed about nearly everything. One favoured one system of coordinates and the other favoured a different system.

Thus, although they both agreed that the distance between a certain pair of adjacent points was so many centimetres, their metric matrices looked quite different. Likewise they agreed that certain curves were geodesics, but the equations of a geodesic looked quite different. They resolved therefore to concentrate on *invariants*, namely, things independent of the choice of coordinates.

So they drew a triangle with geodesic sides and measured the angles, which are indeed invariants. They found that the sum of the three angles was more than two right angles, and they decided that this *excess*, being an invariant, was something to concentrate on. They further realised that the *area* of the triangle was an invariant, and that it might be a good thing to think about the ratio of the excess to the area. They discovered a quite amazing thing about this ratio.

At a certain place on the surface they took ten small triangles, all of the same area but with different shapes, and measured the ten excesses: these ten excesses proved to be all the same. In fact, the ratio of excess to area for small triangles is an invariant, independent of the shape of the triangle and also of the coordinate system.

They decided to call this invariant ratio the *curvature* of the surface and to denote it by the letter B. Now each of the two bugs said in his heart that *his* line-element must contain all the geometry of the surface, and that it must be possible to calculate B in terms of *his* Ax, and yet each bug knew that the other bug favoured a different Ax because he favoured a different coordinate system, and it seemed incredible that the *same* number B should be obtained by two processes starting from such different bases.

But it is in fact so. There is a certain prescription of oper-

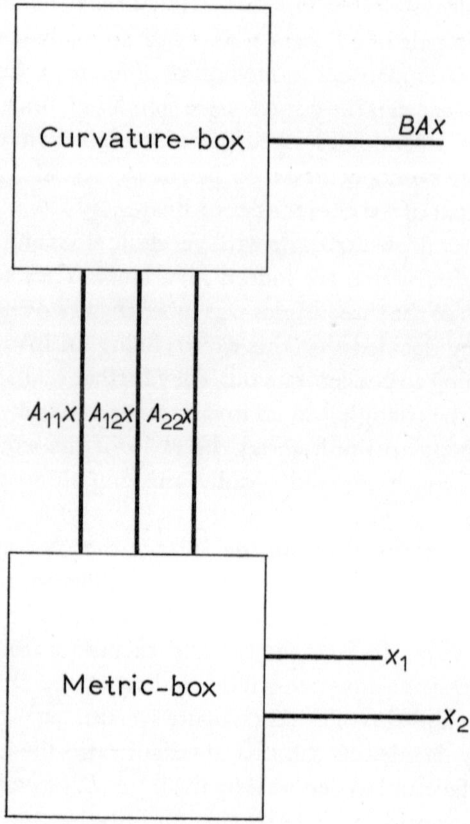

Fig. 10.2

The metric-box and the curvature-box

ations to be performed on the metric matrix Ax, and, if that prescription is followed, the result is independent of the coordinate-system used. Thus there is an *operator* which, without confusion, may be denoted by B, such that BAx is the invariant curvature. There would be no point here in specifying B in detail; it should suffice to say that it tells us to differentiate Ax twice with respect to the coordinates x and then to build up a certain combination of Ax and its first and second derivatives. The essential thing is to observe that this operator B is available once and for all for *all* surfaces and that its discovery (by Gauss) was a discovery of tremendous importance, particularly for us here, since it provides a valuable analogy for relativity.

Fig. 10.2 (which may be compared with p. 44) shows the matter in terms of function-boxes or operator-boxes. As one of the bugs wanders over the surface they inhabit, he is changing his coordinates, that is, he is essentially manipulating the rods labelled x_1 and x_2. This in turn operates through the metric-box on the three rods A_{11}, A_{12} and A_{22}, the machinery inside this box depending on the nature of the surface. Finally, the A-rods operate through the curvature-box on the rod labelled BAx. Unlike the machinery in the metric-box, the machinery in the curvature-box is there once for all, thanks to the genius of Gauss; the two boxes are both operator-boxes, but of very different types.

A word about *curvature*. People are sometimes troubled by the use of the word when there is nothing straight for comparison. That comparison is available if we regard the surface as embedded in Euclidean 3-space – we compare it with a tangent plane. But when we go intrinsic (and we shall have to do that in relativity) there is no 3-space for comparison, and the only thing to do is to accept the word *curvature* as descriptive of properties observed by measurements performed in the surface itself.

It is important to remark that if the curvature vanishes all over the surface, then it is possible to find systems of coordinates for which the line-element takes the very simple form

$$ds^2 = dx_1^2 + dx_2^2 \,.$$

Since this is the line-element of a plane, we then say that the surface is *flat*, although it need not be plane when viewed in 3-space – it might be the surface of a cylinder.

CHAPTER II

The line-element and curvature of a space-time

The heading of this chapter is identical with that of the preceding chapter except for one change: '*surface*' has been changed to '*space-time*'. That is as it should be, because it is a question of setting up an analogy connecting something comparatively simple and familiar with something rather complicated and unfamiliar. So it is right to use the same words, as far as possible, in reference to the two cases. But a curious thing happened. In my first draft of these chapter headings, I put '*a surface*' in the earlier heading, but '*space-time*' without the indefinite article in this one, and I began to wonder why I did that. It is worth talking about. Why do I now insert '*a*' in this heading?

In discussing the geometry of surfaces, I was not thinking of any particular surface, this football or that football, this egg or that egg, this nose or that nose. What I had to say applied to them all, and I considered that 'a surface' conveyed that sense. But when I wrote 'space-time' without the indefinite article, I was falling into the trap I have warned you about so often – I was confusing the R-world and the M_2-world (Pygmalion syndrome!). There is a unique real space-time in which we live, and that deserves to be called 'space-time' without indefinite article, just as John Ferdinand Aloysius Smith should be called by that name without an 'a'. But when it comes to the M_2-world, it is a case of making mathematical models of the R-world, and it may please our fancy to think about some models which differ considerably from the R-world. Indeed it may be said that we never understand anything unless we ask ourselves what it would be like if we changed it a bit. In fact the space-time dealt with in the present chapter may be a space-time rather different from the actual one. It is not a completely free model, however, but subject to the restrictions set out below. So, even at the risk of appearing a bit pedantic, I have stuck in the '*a*' in the chapter heading, but, having made this explanation, I shall use my discretion in what follows.

I have now to submit to you two parallel lists, and I suggest that the proper way to look at this exhibit is to search it for similarities and dissimilarities. The first column should make sense in terms of the discussion in the preceding chapter.

SURFACE	SPACE-TIME
1. Two dimensions.	1. Four dimensions.
2. No specially privileged coordinates x in general.	2. No specially privileged coordinates x in general.
3. ds means distance between adjacent points P, Q.	3. dt means time-interval between adjacent events P, Q in the history of a particle.
4. ds^2 is a quadratic line-element $dx \cdot A \cdot dx$ where A is a symmetric 2×2 metric matrix.	4. dt^2 is a quadratic line-element $dx \cdot A \cdot dx$ where A is a symmetric 4×4 metric matrix.
5. ds^2 is positive-definite.	5. dt^2 is indefinite.

6. Given any point P, it is possible to choose coordinates x so that at P we have $ds^2 = dx_1^2 + dx_2^2$.	6. Given any event P, it is possible to choose coordinates x so that at P we have $dt^2 = -dx_1^2 - dx_2^2 - dx_3^2 + dx_4^2$.
7. There is an operator B (the same for all surfaces) such that BAx is the curvature (a single function).	7. There is an operator B (the same for all space-times) such that BAx is the curvature (a set of 20 functions).
8. If $BAx = 0$ everywhere, it is possible to choose coordinates x such that the line-element has the simple form as in (6) above everywhere (flat surface).	8. If $BAx = 0$ everywhere, it is possible to choose coordinates x such that the line-element has the simple form as in (6) above everywhere (flat space-time).
9. Geodesics satisfy $\delta \int ds = 0$.	9a. Geodesics satisfy $\delta \int dt = 0$.
	9b. There are null geodesics for which $dt = 0$.
10. A string stretched tight on a surface lies in a geodesic.	10a. The world line of a free particle is a geodesic.
	10b. The world line of a photon in vacuo is a null geodesic.

If space-time were nothing but a surface blown up from two dimensions to four, it would be much easier to think about. But it is something very different on account of Items 5 and 6 above. In ordinary geometry we do not meet indefinite line-elements, and this needs some talking about.

But before discussing these Items, let me run over the earlier ones. The first three should be clear enough from earlier chapters, but Item 4 on the right is something new. To see what it means, suppose that a scheme has been set up for attaching four coordinates x to each event in space-time. Imagine then, that an aircraft is flying along, packed with observers each equipped with a clock, all the clocks being synchronised. At a certain instant the aircraft explodes and the observers are hurled out in all directions.

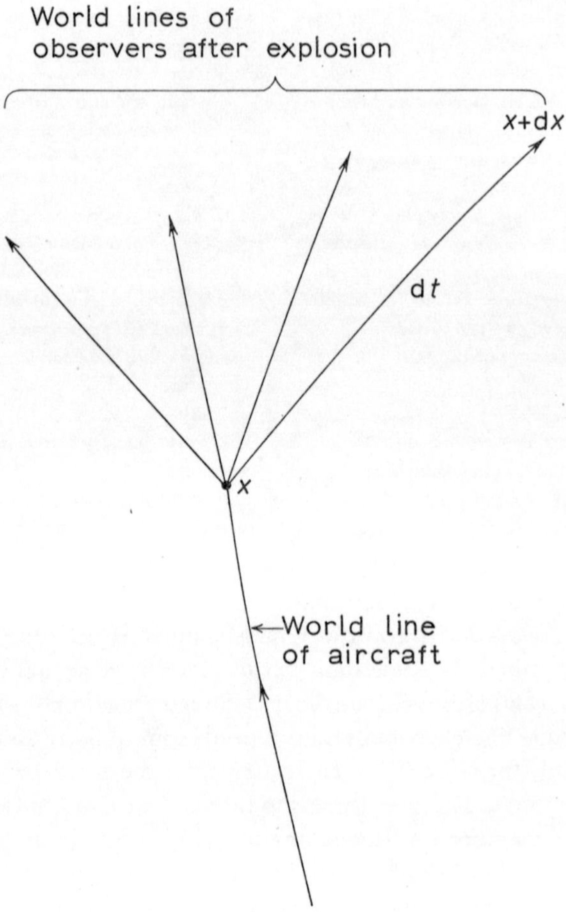

Fig. 11.1

How to find the line-element of space-time experimentally

This is depicted in fig. 11.1 as a space-time diagram, in which we see the history of the aircraft prior to the explosion and some of the subsequent histories of the observers. Only four of these are shown, but you are to imagine that there are many of them.

The coordinates of the explosion are indicated by x (this of course stands for four numbers, x_1, x_2, x_3, x_4), and an event in the subsequent history of one of the observers is $x+dx$. As he flies from the event x to the event $x+dx$ his clock registers a time-interval, say dt.

Now x is a (quadruple) variable, for the explosion might have taken place at any event. Further, dx is also a (quadruple) variable, in view of the multiplicity of observers. Thus dt (or, if you like, dt^2) is a function of the eight variables (x, dx). What sort of a function? That is a question of utmost importance in relativity.

In terms of the fantastic experiment I am describing, it is possible to imagine the many observers getting together afterwards and comparing notebooks. They could then work out what that function is. But of course things don't happen like that. There is a piling-up of ideas, simple ideas leading to more complicated ideas, until the time is ripe for an inspired guess, and this guess is then tested in various ways, usually indirect ways. In this case the inspired guess is that set out in Item 4 above: dt^2 is a quadratic expression in dx with coefficients which are functions of x. It is somewhat tedious to write the whole thing out, but it runs like this*:

$$dt^2 = A_{11}dx_1{}^2 + 2A_{12}dx_1dx_2 + \ldots A_{44}dx_4{}^2 ,$$

* It is usual to write g (for gravity, I presume) where I have written A, but A suits my purposes better and so I shall be unconventional.

the A's being functions of x_1, x_2, x_3, x_4. This I have written compactly as

$$dt^2 = dx \cdot A \cdot dx\,,$$

with A standing for a symmetric 4×4 matrix. There is another compact way of writing this, but for present purposes I shall not bother about it (see chapter 12).

If you count the A's, you will find that, in view of symmetry, there are only 10 of them, although the matrix has 16 entries. I am sorry they are so numerous, but if God chose to make the universe this way, we have to make the best of it.

I would say that, up to this point in the argument, we have met two inspired guesses. First, the guess that time is personal, not universal. Second, the guess that the line-element (which measures that personal time) is a quadratic form in the differentials of the coordinates. By accepting both of these, we have separated the M_2-world of relativity from the M_1-world of Newton in quite a radical way in the matter of concepts. It is a pity that, to drive points home, it is necessary to resort to fantastic thought-experiments – the observers in that exploded aircraft would not have got much useful information from their clocks if some of them had not been shot out with speeds approaching that of light. It would be possible to think up a thought-experiment of a more realistic kind in terms of the interactions of high-energy elementary particles, but the principle of the thing would have been the same.

But now what about the indefiniteness of the line-element, as stated in Item 5? To see what this means, consider an expression

$$dt^2 = dx_1^2 - dx_2^2,$$

like the one already set down on p. 123. This is an *indefinite* form, the word *indefinite* meaning that it can take both positive and negative values. Such a form was quite out of place in the geometry of a surface, but now it fits in nicely. For if we proceed to the form shown in Item 6, we recognise it as indefinite, and the question at once arises: If the displacement dx along the world line of a particle is such that dt^2 is negative (and hence dt imaginary), how does the particle's clock behave? How can it register an imaginary interval of time? The answer is simply that it doesn't. No particle can behave like that. In fact the world lines of particles are not arbitrary curves in space-time – they are such that dt^2 is *positive*.

This calls for an illustration and indeed the space-time diagram (fig. 11.2) is very striking. Although not tied to the exploding aircraft, it is convenient to describe it in those terms. Note first x, the event of explosion, with the world line of one of the particles (or observers) coming out of it, pointing into the future. This world line is representative of all possible world lines, limited by the fact that dt^2 is to be positive and that they should all point into the future. These fill the upper cone, which has the equation $dt^2=0$ and represents the world lines of photons (light-flashes from the explosion). This cone is the *null cone*, or, to be more precise, the *forward* null cone, since the equation $dt^2=0$ defines also a *backward* null cone, which is also shown.

If a particle had been converging on the aircraft, striking it at the instant of the explosion, its world line would be shown inside the backward null cone (it is labelled 'converging world line'). Any space-time direction drawn out from x into the region exterior to the pair of null cones can-

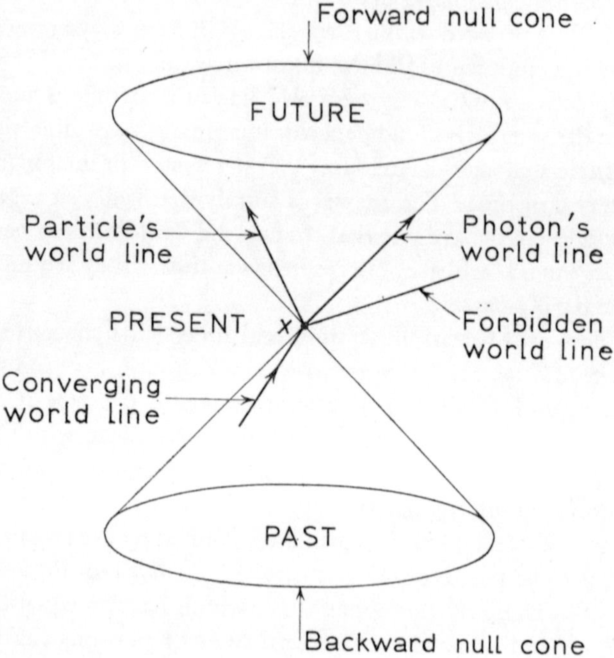

Fig. 11.2
Null cones

not belong to the history of any particle or photon. Although I am trying to avoid the word 'fast', I may say that any such world line would indicate motion 'faster than light'.

This is such a valuable and suggestive picture that one must be excused for cheating. These null cones look like ordinary cones, which are 2-dimensional surfaces in 3-dimensional space; the null cone is actually a 3-dimensional surface in 4-dimensional space-time, and the picture is inadequate. But it is about the best we can do.

The two null cones tell us something about causality. The event x at the common vertex can cause events inside the forward null cone but not elsewhere in space-time. If any effect is caused at x, the cause must lie in the backward null cone. No event outside those two cones can cause any effect at x or be caused by x. The double cone is thus an insulation against causation.

To put this in another way, we are now in a position to use the words *before* and *after* with a little more freedom than earlier. It would be quite correct to say that any event lying in the forward null cone is after the event x at the vertex. For if any particle included that event and the event x in its history then that event would come after x in the experience of that particle – that is a matter of the particle's personal clock. Equivalently one may say that x is before such an event.

Likewise one may say that all events lying in the backward null cone were before x.

But what about events lying, not in either null cone, but in the region outside it? It is impossible that any particle could include in its history both such an event and the event x itself. There is no particle which could establish a time-order, and so the idea of a time-order disappears in such a case. Events outside the null cone are neither future or past. It is right to attach the words FUTURE and PAST to the two pieces of the null cone, but one is a bit at a loss what word to attach to the rest. I have put the word PRESENT in fig. 11.2, but with some misgivings. Would you say you were 'present' at a celebration if it was impossible to send a message of congratulation or receive messages from those celebrating? It is true that you might have sent a message before your world line left the past, and you could receive one when your world line had passed into the future, but, in between, you would be incommunicado. Perhaps you should delete PRESENT and substitute INCOMMUNICADO.

Having thus got through Items 4 and 5, let us turn back to p. 131 and take a look at Item 6. On the left hand side it tells us that, given any point P on a surface, we can find coordinates so that

$$ds^2 = dx_1^2 + dx_2^2$$

at P. Mind you, at P, not everywhere, for, if everywhere, then the surface would be flat. What we seem to be saying is this: a very small part of a surface may be regarded as plane. In other words, when laying out flower-beds, it is unnecessary to allow for the curvature of the earth. Or is it unnecessary? It depends on the size of your garden and the precision to which you work. If you were neurotically tidy, you might be upset by the impossibility of making a flower-bed in the form of a square with sides a mile long, for, strictly speaking, you cannot draw a square on the surface of a sphere. It is clear then that Item 6 calls for some caution. It is sometimes stated thus: a surface is elementarily flat. But this is nonsense, because, no matter how we choose the coordinates, the curvature of the surface remains – it is invariant.

As worded on p. 131, Item 6 is technically correct in both columns, left and right, but it is a matter to be handled with care when we seek physical interpretations. It is commonly stated that space-time is elementarily flat, but I would advise you to accept that phraseology with caution – it might cause confusion in cultivating your four-dimensional garden in space-time. Let us leave Item 6 with that. In a way it is important, but I don't think I shall have to refer to it again.

Item 7 deals with curvature, and I have dealt fairly fully with that concept for a surface. You will recall that the curvature is one number at each point on a surface. If bugs on the surface know what its curvature is, they can plan their lives accordingly. Just as the pilots of two aircraft, starting from the North Pole simultaneously and flying due south at the same speed, know that they will meet at the South Pole, so could two bugs on a surface estimate the deviation or convergence of geodesic paths. All they need to know is the curvature BAx at each point x on the surface, B being the curvature operator.

Riemann discovered that these ideas could be carried over into a space of any number of dimensions, but in a much more complicated form when you go into details. What he did applies to 4-dimensional space-time. There is an operator B which, when applied to the metric Ax, gives a set of 20 functions of x, viz. BAx in the notation I am using. This set of 20 quantities is called the curvature tensor, or, equivalently, the Riemann tensor (in the next chapter I shall talk about tensors). To us (bugs in space-time) this curvature may be interpreted in terms of the deviation of geodesics representing the histories of free particles and photons (see Items 9 and 10 on p. 131). Since celestial mechanics is concerned with free particles (planets) and the light by which we observe them, the curvature of space-time is a very real thing in the physical world.

I am not going to write down the formula for the curvature tensor. It is complicated. The important thing is that this operator B which turns the metric Ax into the curvature BAx is a quite definite set of instructions – you are to differentiate Ax twice and make a certain algebraic pudding out of the first and second derivatives and Ax itself. If you like, you can think of function-boxes as on p. 126, but now there are four x-rods, ten Ax rods and twenty BAx rods.

What is gravitation? It is the curvature tensor. That is the correct answer as far as the M_2-world of Einstein is concerned. If (see Item 8 on p. 131) the curvature tensor vanishes, space-time is flat. It is the curvature that constitutes the gravitational field.

The general theory of relativity deals with curved space-time and the special theory of relativity with flat space-time. Thus the general theory deals with gravitation and the special theory with physical phenomena in which gravitation is so unimportant that we can forget about it. Far more people are acquainted with the special theory than with the general – the mathematics of the special theory is much easier and so many more problems can be solved. I have chosen the hard course of talking first about the general theory, and this I did for the reason stated earlier – it is easier to break with Newtonian concepts if we plunge at once into the general theory. But a price has to be paid. I could not inflict on you the intricate details of formulae and so, as I realise only too well, I have been peopling your minds with ghosts of which the only recognisable features are crudely numerical. The curvature tensor has twenty components, but that information would not enable you to recognise a curvature tensor if you met it in the street. You do, every day of your life, unless you have discovered some secret way to make yourself immune to gravitation.

Having gone so far with the general theory, I must carry on a bit further. So I invite your patience for a chapter on tensors and one on the field equations of Einstein.

CHAPTER 12

Tensors

In this book I have avoided the use of formulae as far as possible, but, when it comes to tensors, I cannot keep this up. To say what has to be said in ordinary words would be insufferably tedious. But I shall keep the mathematical formalism to a minimum.

Suppose we have a set of sixteen quantities which we wish to exhibit. They may be set out in the form of a matrix:

$$\begin{pmatrix} X_{11} & X_{12} & X_{13} & X_{14} \\ X_{21} & X_{22} & X_{23} & X_{24} \\ X_{31} & X_{32} & X_{33} & X_{34} \\ X_{41} & X_{42} & X_{43} & X_{44} \end{pmatrix}$$

But that is tedious. The whole set may be denoted by a single symbol X; but for present purposes it is better to write X_{ij}, with the understanding that i and j stand for numbers in the range 1, 2, 3, 4.

I use the range 1, 2, 3, 4 because it is the range needed in relativity, in which the number *four* is, so to speak, a magic number. But most of what I have to say would apply equally well to a range 1, 2, ...n.

Since, for our purposes, the range is always 1, 2, 3, 4, it is unnecessary to go on repeating the instructions: the symbol X_{ij} implies the range without further comment.

Suppose now that X_{ij} are functions of space-time coordinates x_i; this is indicated by writing $X_{ij}(x)$. Let y_i be another system of space-time coordinates. This means that the x's are functions of the y's and vice-versa, and we may speak of the partial derivatives $\partial x_i/\partial y_j$ (sixteen of these) and $\partial y_i/\partial x_j$ (sixteen of these).

Take the sixteen functions $X_{ij}(x)$ and substitute for the x's in terms of the y's; this gives sixteen functions of the y's which we shall denote by $Y_{ij}(y)$.

In general there is no interesting simple relationship between the X's and the Y's, but it may happen that there is a relationship of the following form:

$$X_{ij}(x) = \Sigma Y_{km}(y) \frac{\partial y_k}{\partial x_i} \frac{\partial y_m}{\partial x_j}, \tag{12.1}$$

where Σ means that we are to give to the subscripts k and m the values 1, 2, 3, 4 and add together all the terms so obtained. If the relationship holds, we say that the X's form a *tensor*, and the above formula is a *tensor transformation*.

There is an awful lot packed into (12.1). There are sixteen equations, and each equation has sixteen terms on the right-hand side. But (12.1) says all this – it is a triumph of mathematical laziness. But Einstein was lazier still, for he saw that whenever such a formula called for summation over a suffix (k and m in (12.1)), that subscript always appeared *twice*. Thus the sign of summation (Σ) is really unnecessary – scratch it out and write the law of tensor transformation as follows:

$$X_{ij}(x) = Y_{km}(y) \frac{\partial y_k}{\partial x_i} \frac{\partial y_m}{\partial x_j}. \tag{12.2}$$

This is called the summation convention: *when a subscript is repeated in a product, sum with respect to it over the range of numbers* 1, 2, 3, 4. Equation (12.2) says precisely what (12.1) says, but it says it a little more compactly.

A tensor transformation has two important properties, and it is these properties which make tensors such useful (indeed essential) tools in relativity.

The first property is that, if a tensor vanishes in one coordinate system, then it vanishes in all coordinate systems. This is obvious – if all the Y_{ij} are zero, then all the X_{ij} are zero.

The second property is that the sum of two tensors is itself a tensor. To see this, write down a second equation like (12.2):

$$A_{ij}(x) = B_{km}(y) \frac{\partial y_k}{\partial x_i} \frac{\partial y_m}{\partial x_j}.$$

Now add this to (12.2) and get

$$X_{ij}(x) + A_{ij}(x) = [Y_{km}(y) + B_{km}(y)] \frac{\partial y_k}{\partial x_i} \frac{\partial y_m}{\partial x_j}.$$

This is a tensor transformation – the sum $X_{ij}(x) + A_{ij}(x)$ is a tensor.

I have been at pains to emphasise that in general there are no specially privileged coordinates in space-time. That means that, if we have chosen to use some coordinates (say x), we must be prepared to change to other coordinates (say x'), and of course we need to know what happens to our formulae under this transformation of coordinates. It is a question of book-keeping, with some analogy to what happens when an English-speaking country decimalises. The trick in relativity is to work with tensors, for then we have the tensor transformation (12.2) – it may look formidable, but it is nothing to what we might have without a good notation.

But where do we start putting tensors into relativity? The answer is that we have started already. You will recall that in the preceding chapter we had a line-element of space-time, which I wrote in the brief form

$$dt^2 = dx \cdot A \cdot dx \,.$$

Here A was the metric matrix; we may write A_{ij} for it, and express the line-element as

$$dt^2 = A_{ij} dx_i dx_j$$

(remember the summation convention!). Now dt is a measure of time by means of a clock carried by a particle, being in fact the time-interval recorded as the clock passes from the event x_i to the event $x_i + dx_i$. This time-interval must remain the same if we change the coordinate system, and so we have

$$dt^2 = A'_{ij} dx'_i dx'_j = A_{ij} dx_i dx_j \,, \tag{12.3}$$

where x'_i are the new coordinates, and A'_{ij} the metric matrix for that system of coordinates. But i and j are 'dummy' subscripts (to be summed); let us change them to k and m in the last term in (12.3), so that the equation reads

$$A'_{ij} dx'_i dx'_j = A_{km} dx_k dx_m \,.$$

Now

$$dx_k = dx'_i \frac{\partial x_k}{\partial x'_i}, \qquad dx_m = dx'_j \frac{\partial x_m}{\partial x'_j},$$

and so our equation becomes

$$A'_{ij} dx'_i dx'_j = A_{km} \frac{\partial x_k}{\partial x'_i} \frac{\partial x_m}{\partial x'_j} dx'_i dx'_j.$$

Since dx'_i is an arbitrary displacement in space-time, this is an identity in the differentials, and hence it follows (since the metric matrix is symmetric) that

$$A'_{ij}=A_{km}\frac{\partial x_k}{\partial x'_i}\frac{\partial x_m}{\partial x'_j}. \tag{12.4}$$

This is a tensor transformation and so we may call A_{ij} the *metric tensor*. It is often called the *fundamental tensor*, since it is the building-block from which other tensors are constructed.

In particular we had an operator B such that BAx describes the curvature of space-time. Actually this is a tensor, but of a more complicated type, having four subscripts instead of two: we write it B_{ijkm}. It looks as if there were 256 components here, but in fact they are interrelated, so that they boil down to 20.

It is true for all tensors that, if they vanish for one coordinate system, they vanish for all coordinate systems. I remarked earlier that $BAx=0$ is the criterion for flatness of space-time. This would have been a silly thing to say if we were not dealing with a tensor, for we expect flatness to be a property independent of the coordinate system we use. But we are and it is.

CHAPTER 13

The field equations

My plan in this book has been to start with the general and proceed to the particular. To recapitulate briefly, I began with the four-dimensional space-time of events, with world lines representing the histories of particles, and, for events occurring on one world line, a time-order (before-and-after). In all that there was nothing foreign to Newtonian ideas. The approach was unusual but unobjectionable from the Newtonian standpoint – the work of a crank who refused to make use of such common words as *speed* and *distance*, but who otherwise expressed himself sensibly.

The crunch came when I presented time as a personal thing, to be measured by atomic clocks carried along by the observers. With this demotion of Newton's absolute time, we were definitely launched on something new. But there remained many details to be filled in and at this point the simple intuitions hitherto available had to give place to mathematical formulae. My hardest problem has been to find a middle road through those formulae. If I were to present them in all detail, you would not see the wood for the trees, unless you were a professional relativist, or so it seemed to me, and I have attempted a sort of thumb-nail sketch.

I did however deal honestly with the metric tensor, that set of ten functions of the coordinates involved in the line-element dt^2, dt being the clock-measure of a small time-interval. It was a relief from formalism to present the null cones depicted on p. 136, and so interpret the statement that no particle can travel as fast as light without saying what *fast* means.

I skated lightly over the complexity of the curvature tensor B, that set of twenty functions of the coordinates which has the duty of distinguishing curved space-time from flat space-time, in much the same way as the Gaussian curvature of a surface distinguishes it from a plane.

But this will never do. I have been forgetting the title of this book and the aim I set before me – to talk, not to fall into didactic boredom. I can almost hear you yawning. You are tired of these faceless tensors. You would rather be hearing about reality, and what I seem to be offering you is a world of vague mathematical symbols. The time has come for me to make a confession about that Pygmalion syndrome of which I spoke disparagingly in chapter 1. I warned you of the danger of confusing the real R-world with the model M-worlds constructed in order to understand the R-world. My confession is that I myself suffer from that disease. Mathematicians may enjoy pushing round complicated formulae – they do – but the call of nature is always there – physics keeps breaking in. So, as I sit here at my table grinding out words, I ask myself what are the most real things in this room, and, when I say *real*, I mean deep and fundamental. There are books and a type-writer and all that, but they are, in a sense trivial. What, I ask myself, is the real structure underlying all these trivialities, and the answer is – the metric tensor!

Yes, indeed, those ten little fellows are dancing in the tobacco smoke. They are the ultimate reality, but elusive in the sense that, as I strike this key of the typewriter, they do indeed exist, but the values they have depend on what coordinates I use, and I have never got round to setting up a system of space-time coordinates in my room because I know that all systems of coordinates should be treated with even-handed justice – there is no privileged class of coordinates.

Let us then surrender to the Pygmalion syndrome, merging the R-world and the M_2-world of relativity. In this spirit, the metric tensor becomes as real as the nose on your face and much more fundamental. It is measurable, in principle, by means of a clock. Admit, too, as a reality the more elusive curvature tensor with its twenty components. It also is measurable, in principle, by observing how the tracks of two free particles diverge or converge as they fly along. Do not accuse me of talking through my hat. These are ultimate realities of nature, and if they do not intrude more obviously on our senses, it is because our senses are too blunt and most of our exploring apparatus is too blunt also. But it is getting more acute all the time, and the remote mystery of today becomes the reality of tomorrow. It is not so many years ago since I instanced the other side of the moon as one of those things men would never see – how foolish I feel now!

The time has come to add the final storey to the relativistic structure by saying how the metric tensor is connected with the matter in the universe. In view of the freedom of choice of coordinates, this relationship must be a tensor relationship. So we are to seek two tensors and equate them. One tensor is to be constructed out of the metric tensor (which I have called Ax, this notation indicating that it is a function of the space-time coordinates x). The other tensor is to be constructed so as to represent matter (I shall call it Mx).

We have already constructed a tensor out of Ax, namely the curvature tensor BAx with twenty components, but for various reasons that will not do. One reason is that the curvature tensor BAx has twenty components while it seems that the matter tensor Mx has only ten, and so you cannot equate them. However Einstein found that there was another more suitable tensor available, closely related to BAx but with only ten components. It is customary to use the letter G for it (perhaps G for gravity) and I shall do this. You may regard G as an operator; when you apply it to any metric tensor Ax, you get a tensor GAx (the Einstein tensor), and the *field equations* read

$$GAx = Mx\,. \tag{13.1}$$

In a vacuum, where there is no matter, these equations read

$$GAx = 0\,. \tag{13.2}$$

These are the equations which govern gravitational fields. When written out, they consist of ten partial differential equations of a very difficult type, perhaps the most difficult equations occurring in the whole of physics, and for that reason a great challenge to mathematicians. Very few interesting simple solutions are known, but at least the equations can be solved for a universe in which there is only one massive body which is spherically symmetric, and that fits the solar system fairly well. Although astronomers continue to use Newtonian mechanics for most of their calculations, the general feeling seems to be that (in Pygmalion language) Einstein's equations are correct, but terribly hard to solve.

The discovery of the operator G was the grand climax in Einstein's theory of relativity. Without it, we would have no theory of gravitation with which to replace Newton's theory. I refrain from writing down the complicated formula for G, and shall merely say that it involves definite instructions what to do with the metric tensor Ax: differentiate it twice and follow a certain recipe in combining Ax itself with its first and second derivatives. That is very like what I said about B – the recipe is different, that is all.

I must say a few words about the matter tensor Mx. It has ten components, which we may write M_{ij} if we like. One of these components represents *density*, and that reminds us of Newton for whom the *mass* of a body (or equivalently its density) was the thing that mattered when it came to gravitational attraction according to the familiar inverse-square law. But density is only one component of M_{ij}. There are nine more to account for, but if we choose the coordinates properly we can reduce those nine to six at any specified event. Those six represent the *stress* (pressure, tension and shearing stress) in the body, and in this Einstein deviates from Newton for the latter attributed no gravitational effects whatever to stress.

One of the things that bedevils attempts to make relativity realistic is the relative *smallness* of certain quantities. When expressed in comparable units, stress is much smaller than density, and to our blunt senses it may seem that it is a needless refinement to bring stress into the picture at all. But if we allow approximations of that sort to intrude, we are likely to find ourselves in a confused position, halfway between Newton and Einstein. Therefore, small as the effect may be, we must accept the fact that stress influences gravity.

The only way to understand all this is to pretend that our faculties and measuring instruments are much more refined than they are. I propose then to invite you into my room while I conduct some impossibly refined experiments.

The first thing to do is to don oxygen masks and have all the air evacuated from the room, for, unless we do this, we shall be annoyed by the bombardment of molecules. Let us leave the ordinary furniture in the room, but introduce some apparatus of a conceptually simple nature. We shall need some particles with clocks in them and also some mechanism in them for emitting little flashes of light and receiving such flashes.

The object of the exercise is to explore the gravitational field in this room of mine, and I strongly expect that you imagine that you know all about it before we begin. For you will tell me that every object in the room is attracted by the earth in accordance with Newton's law of the inverse square, and that the objects attract one another according to that same law. You speak of *forces* and *accelerations*, but, frankly, I don't know what you mean. I have long since abandoned the attempt to attach relativistically viable meanings to those words, and I know perfectly well that if you start to explain them you will use other terms equally inexplicable except in terms of Newtonian concepts. So don't pretend you know all about it.

To cut short discussion, I perform an experiment. I set two of the particles on the edge of my table and give them a flick with my finger. Away they go, with tiny flashes of light being exchanged between them. From the readings of the clocks on the particles we can follow the behaviour of their relative distance, as discussed on p. 113 (see fig. 9.3). Is this distance proportional to the time-interval from the flicking-off? If it is, and if this holds for all such experiments, then I say that there is no gravitational field present. This might very well be the case if my room had been transported to remote space, far outside the solar system, but I don't believe that it would happen in this very room here. On the contrary, I believe that the distance between the particles would *not* be proportional to the time elapsed, and this lack of proportionality would be evidence of the existence of a gravitational field. The two trajectories would, so to speak, bend in or bend out relative to one another, and the amount of this bending would be a measure of the strength of the gravitational field.

If you ask me what a gravitational field is, I can only repeat what I said earlier – it is the curvature tensor, the thing I have called B. To measure B in my room we would have to carry out a number of experiments with particles like the one I have described, but we would also have to choose some system of coordinates, because the values of the twenty components of B jump about when we change coordinates. In fact, it is very hard to pin down the term *strength* in relation to a gravitational field, but in general terms we might say that the field is strong when the components of B are large and weak when they are small, but even this is not clear since the values depend not only on the coordinates used, but also on the units employed.

The general conclusion is that gravitation is curvature of space-time, and this curvature is described by the tensor B but it is hard to make the same sort of simple statements as are made in Newtonian theory. It is a pity, but in this Pygmalion mood I claim that it is better to make true statements, even if they are complicated, than false statements, which have no merit except simplicity.

Let it be granted then that, in this room of mine, there is some curvature tensor B, capable of being measured with sufficiently refined instruments. This tensor B is, through the field equations, determined by the energy tensor Mx, or so it appears. The most important part of Mx is due to the earth, but there are contributions from my body, my table, my chair and so on. Thus, if I move the furniture round, the gravitational field will change. But even if I do not move the furniture, the dependence of the gravitational field on stress means that I change the field by merely *leaning* on my table, for that will set up stress in the table and in my arm.

The very fact that the field equations are so hard to solve makes one all the more inclined not to try to do that, but to play round with them. It is a mathematical tradition that, in any equation, we put unknowns on the left and knowns on the right. In what has just been said about leaning on the table, it was tacitly assumed that Mx was given and the field to be found. The way in which I wrote the field equations suggested that. But there is really no reason why the equation should not be written the other way round:

$$Mx = GAx \,. \tag{13.3}$$

It is of course precisely the same equation as (13.1), but the psychological setting is different.

This new form of the field equations suggests that we might do a bit of universe-building. Start with any metric tensor Ax, subject only to the condition that it should give an indefinite line-element of the right sort, which is a pretty weak restriction. Having written down the ten functions Ax, you now apply the operator G and work out the ten functions GAx. This is a purely routine operation, easy to carry out. Then use (13.3) to *read off* Mx, and you have a world satisfying Einstein's field equations! In this way you might fill a book with imaginary universes.

But there is a snag in it. The universes you get by this simple process are very likely to suffer from a defect which makes them unrealistic – the density turns out to be negative somewhere in the universe. If you are concerned only with playing with symbols, this may not seem important, but it is not a thing physicists like. Perhaps they are narrow-minded, but the fact is that negative densities are unacceptable and this simple way of building universes is not much good after all. Nevertheless, curiously enough, it was when I first wrote the equations wrong way round as in (13.3) that I began to understand what the field equations really meant – the suggestion was due to Professor W.H. McCrea, and I am very grateful to him.

As every student of mathematics knows, it is a useful thing to check your results. If you are asked to solve some equation and succeed in doing so, it is wise to substitute your answer in the equation and see if it is really satisfied. Perhaps you don't claim to have more than an approximation to the solution (as when you say that $x=1.414$ satisfies $x^2=2$), and then you can assess the magnitude of the error by substitution. The field equations do play that role when written in the form (13.3).

For suppose you are trying to find the metric tensor Ax

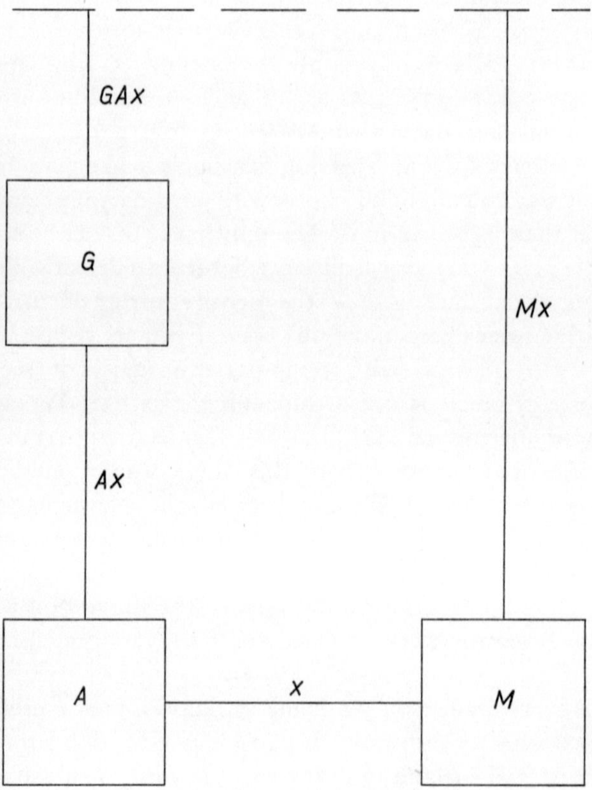

Fig. 13.1
Einstein's field equations

in the region of the solar system, outside the sun or any of the planets. By some tedious involved method of approximation, you arrive at some result for Ax, but you are not sure whether it makes sense or not. So you apply the operator G and work out GAx, the Einstein tensor. Now since interplanetary space is a vacuum, you should have $Mx=0$ there, and if GAx fails to be zero, you know there is something wrong. In the lingo of relativists, there is 'dirt' where there should be vacuum. By (13.3) you can calculate that dirt and then make up your mind whether it is so small as to be negligible. For a number of years now I have been working with this dirt, trying to get rid of it, and so Einstein's field equations sometimes appear before my mind's eye in the form (13.1) and sometimes in the form (13.3).

It might help to exhibit the field equations in terms of function-boxes or operator-boxes, as in fig. 13.1. The rod labelled x is really four rods representing the four coordinates in space-time. On its left, we have the box which converts x into Ax, the metric tensor, represented by one rod, but really ten rods. Above that we have the magical box of Einstein, the operator which changes Ax into the Einstein tensor GAx, and here again there are really ten rods sticking up. On the right we have the box M converting x into the matter tensor Mx, and here too there should be ten rods sticking up. In a crude symbolism, the equations $GAx=Mx$ are indicated by making the rods on the left rise to the same height as those on the right. Such a diagram is not to be taken too seriously, but it does, I think, show up the connections between things.

CHAPTER 14

From general to special

Laying aside as irrelevant all attempts to find literal interpretations for the word *relativity* (and I feel that the great majority of relativists would support me in this), let me repeat that the general theory of relativity is Einstein's theory of gravitation, while the special theory excludes gravitation. This means that if you want to deal properly with any gravitational phenomenon, you must invoke the general theory, but if gravitation is insignificant in relation to the phenomenon under discussion, then you turn to the special theory instead. The special theory is (as its name implies) a special case of the general. It deals with flat space-time, and so there is no more talk of the curvature tensor B (it vanishes) or of Einstein's tensor G (it vanishes too). This makes for simplicity, but this simplicity encourages us to go into more details and these bring in their own complications.

I have now to stop harping on the fact that there are no privileged coordinates in space-time, because in special relativity there is indeed a privileged class of coordinates, namely those for which the line-element takes the simple

Minkowskian* form

$$dt^2 = -dx_1^2 - dx_2^2 - dx_3^2 + dx_4^2 \,. \tag{14.1}$$

Such coordinates are not unique by any means, any more than rectangular Cartesian coordinates in a plane are unique. In the plane you are free to translate and rotate the axes of coordinates, and a similar transformation (the Lorentz transformation) is available in flat space-time, leaving the line-element unchanged. The above simple form holds for all such coordinates. From a mathematical standpoint, the special theory of relativity is essentially the theory of Lorentz transformations. This is a rich subject with many detailed results – special relativity is an extensive subject in itself.

As I have said more than once, I decided to put general relativity first because it offers a better opportunity for shaking off Newtonian preconceptions, and I hope I have succeeded reasonably well in this endeavour. In passing to the special theory, you do not have to change the concepts already encountered, nor do you have to cope with new ones. As far as concepts are concerned, the situation is little changed. But we still have with us the two old bugbears – four-dimensionality and an indefinite line-element. Geodesics, however, are much easier to think about, for in flat space-time they are like straight lines in a plane rather than curves on a surface – two of them cannot intersect in more than one event.

To get back to the privileged coordinates for which the line-element has the simple form (14.1), such coordinates remain rather nebulous until we decide how they are to be realised in nature – or in the M_2-world of special relativity,

* It was Minkowski, not Einstein, who invented the concept of space-time.

as we should say if we are to keep free of the Pygmalion syndrome. How are such coordinates to be realised?

The usual answer to this question runs somewhat as follows: Take a rigid body. In it take three rectangular axes, and let x_1, x_2, x_3 be the space-coordinates of an event relative to those axes. Supply the rigid body with a system of synchronised clocks, and let x_4 be the time of the event as recorded by the clock at which the event occurs. Then these four x's are space-time coordinates for which the line-element is as in (14.1), provided the unit of time is suitably related to the unit of length and provided the body does not rotate and is so far away from massive bodies that gravitational effects may be disregarded.

There are parts of that statement which jar on me, and I would like to modify it.

First, it is very dangerous to speak of *rigid* bodies, even in special relativity; the concept of rigidity is unacceptable. So, instead of taking a rigid body, let me start with clocks. Let us think of a tremendous number of clocks, each of them moving freely (no forces acting on them) and all ticking away. But we do not leave the motion of these clocks arbitrary – if we did, this cloud of clocks would dissipate itself. Referring to p. 113, we recall that any one of the clocks can measure the relative distance of any event from it. Thus a clock A can keep a continuous record of the distance of another clock B from it, and, unless precautions are taken, this distance will change continuously. However, it follows from the geometry of flat space-time, combined with the fact that each clock pursues a geodesic, that if they are given suitable motions to begin with, then all the mutual relative distances (the D of p. 113) will remain constant. Let this be done. Then we may say that, as far as these distances are concerned, the cloud of clocks behaves as we would expect a rigid body to behave – the distances do not wobble about.

Referring again to p. 113, we note that a clock A can keep a continuous record of the relative times T at which events occur at another clock B. Appealing again to the geometry of flat space-time and the geodesic character of the world lines, we find that the zero-readings of the clocks can be set so that the relative time of an event on clock B coincides with the time B is reading at that event. This means that any event, wherever and whenever it occurs, can receive a time-coordinate x_4, unequivocally, either by reading the clock at which the event occurs or by noting its relative time as recorded by any of the clocks, for we have arranged that these two numbers shall be the same.

So now we have a cloud of clocks moving rigidly and they are synchronised in the above manner. We have attached a coordinate x_4 to an event. What about the other three coordinates?

The cloud of clocks form a *frame of reference*. Relative to it, we have succeeded in separating time from space. In the space of the clocks, there is a distance D between any two clocks, an unchanging distance, which is just what we have in ordinary Euclidean geometry of three dimensions. It is a consequence of the flatness of space-time (and the way in which we made the clocks move) that this is in fact a Euclidean 3-space. We can accordingly take rectangular Cartesian coordinates in it, and thus attach the coordinates x_1, x_2, x_3 to any event.

In this way all the four coordinates have been assigned and the line-element is as in (14.1'). In reaching this result I have suppressed the mathematical deductions underlying it. These can easily be supplied if needed. What I wanted to do was, not to prove things mathematically, but to show that the concepts involved are all as clear as they should be.

If we use another frame of reference (another clock-cloud) we shall again split space-time into space and time, but it will be a different splitting. The values of the four coordinates assigned to an event will not be the same as before – the two sets of coordinates will be related to one another by a Lorentz transformation.

In talking about two frames of reference, we may refer to them as S and S′, and we may use those letters to refer also to *observers* attached to the two frames of reference. These observers are merely agents who record the readings in the two clock-clouds. To any event, S attaches four coordinates x_i and to the same event S′ attaches four coordinates x'_i. The problem is to express one set of four coordinates in terms of the other four, and the key to the situation is the invariance of dt, the time-interval between two events in the history of a particle, as recorded by its own clock. This particle may be one of the clocks in S or one of the clocks in S′, but in general it is neither. The essential formula is

$$\begin{aligned} dt^2 &= -dx_1^2 - dx_2^2 - dx_3^2 + dx_4^2 \\ &= -dx_1'^2 - dx_2'^2 - dx_3'^2 + dx_4'^2. \end{aligned} \qquad (14.2)$$

The mathematics of the Lorentz transformation is greatly simplified if S and S′ cooperate. Without changing the frame of reference, S can make certain changes in the way he assigns coordinates to events. In fact, he can make three adjustments: (i) he can adjust the zero-reading of his synchronised clocks, (ii) he can adjust the origin of the three space-coordinates, and (iii) he can rotate those space-axes. S′ can make the same three adjustments. It is easy to see that if either S or S′ makes suitable adjustments (i) and (ii), they can bring it about that S and S′ both assign zero values of all four coordinates to some chosen event; that is, there is some event for which

$$x_i=0 \quad \text{and} \quad x'_i=0. \tag{14.3}$$

The cooperation of type (iii) is more subtle, and I shall not attempt to prove the following statement: S and S′ can rotate their space-axes in such a way that $x_2=x'_2$ and $x_3=x'_3$ *for all events.*

The identity (14.2) may now be written

$$-\mathrm{d}x_1^2+\mathrm{d}x_4^2=-\mathrm{d}x_1'^2+\mathrm{d}x_4'^2. \tag{14.4}$$

To see what this implies, it is convenient to define four new variables

$$\begin{aligned} y&=x_1+x_4, \qquad & y'&=x'_1+x'_4\,, \\ z&=x_1-x_4, \qquad & z'&=x'_1-x'_4\,. \end{aligned} \tag{14.5}$$

Then S may use (y, z) instead of (x_1, x_4) and S′ may use (y', z') instead of (x'_1, x'_4). In terms of these new variables the identity (14.4) becomes

$$\mathrm{d}y\,\mathrm{d}z=\mathrm{d}y'\,\mathrm{d}z', \tag{14.6}$$

and it is easy to guess how to solve this. We put

$$y'=ay, \qquad z'=a^{-1}z \tag{14.7}$$

where a is some constant. *This is the Lorentz transformation in its simplest mathematical form.*

To interpret this transformation in a more physical way, we go back to (14.5) and recognise that (14.7) gives

$$x'_1+x'_4=a(x_1+x_4), \qquad x'_1-x'_4=a^{-1}(x_1-x_4)\,,$$

and hence, by addition and subtraction,

$$x'_1=\tfrac{1}{2}(a+a^{-1})x_1+\tfrac{1}{2}(a-a^{-1})x_4\,,$$
$$x'_4=\tfrac{1}{2}(a-a^{-1})x_1+\tfrac{1}{2}(a+a^{-1})x_4\,. \qquad (14.8)$$

Here we have a second form of Lorentz transformation, a little more physical than (14.7). To make it still more physical, let us define new constants γ and v by

$$\gamma=\tfrac{1}{2}(a+a^{-1}), \qquad \gamma v=-\tfrac{1}{2}(a-a^{-1})\,,$$

noting the identity

$$\gamma^2-\gamma^2v^2=1\,,$$

so that

$$\gamma=(1-v^2)^{-\frac{1}{2}}. \qquad (14.9)$$

Now we have a neat physical form for the Lorentz transformation:

$$x'_1=\gamma(x_1-vx_4), \qquad x'_4=\gamma(-vx_1+x_4)\,, \qquad (14.10)$$

together with $x'_2=x_2$, $x'_3=x_3$.

I call this *physical*, but this is a sop to Newtonian ideas, in which *velocity* is basic. Suppose we follow a particle of the S′-cloud. Its coordinate x'_1 is fixed, and so $dx'_1=0$. Then (14.10) gives

$$dx_1/dx_4=v\,. \qquad (14.11)$$

Thus we may say that the constant v is the velocity of S′ relative to S.

It is easy to solve (14.10) for the unprimed x's in terms of the primed. In view of (14.9), we get the remarkable result:

$$x_1=\gamma(x'_1+vx'_4), \qquad x_4=\gamma(vx'_1+x'_4)\,, \qquad (14.12)$$

which is the same as (14.10) except for the interchange of primed and unprimed and the substitution of v for $-v$. Thus if we now follow a particle of the S-cloud (so that $dx_1=0$), we get

$$dx'_1/dx'_4=-v\,, \tag{14.13}$$

and we may say that the velocity of S relative to S′ is $-v$.

As excuse for the preceding argument, I thought it would be interesting to treat the simple formula (14.7) as basic and make deductions from it. But, if this bores you, you can immediately *verify* that the final formulae (14.10) and (14.12) satisfy the basic identity (14.4), remembering the definition of γ in (14.9).

Note that, if $v=0$, then we get $\gamma=1$ and $x'_1=x_1$ and $x'_4=x_4$, so that the Lorentz transformation is an identity. Further note that, to make γ real, we must take v less than unity in absolute value; in fact $v=1$ is the velocity of light and the relative velocity of the two clouds must be less than this.

This is the place to talk about the apparent contraction of a moving body and the apparent slowing-down of a moving clock.

Let us then concentrate on two adjacent S′-clocks with coordinates x'_1 and $x'_1+dx'_1$, both constants. Let us further concentrate on two *events* in the histories of those two clocks, namely, events which are simultaneous as judged by S so that $dx_4=0$ when we compare those two events. Then, by the first of (14.10) we have

$$dx'_1=\gamma\,dx_1\,.$$

This may be written

$$dx_1/dx'_1=\gamma^{-1}=(1-v^2)^{\frac{1}{2}}. \tag{14.14}$$

This is clearly less than unity. Now dx_1 is the distance between the two S′-clocks as observed by S and dx'_1 their distance as observed by S′; the former is less than the latter in the indicated ratio. This is the *apparent contraction* of a moving body (or, as I would prefer to put it, a moving clock-cloud).

Now we concentrate on two events in the history of an S′-clock. For those two events we have $dx'_1=0$ (the clock retains its S′-position) and between them there is an S′-time-interval dx'_4. Let dx_4 be the S-time-interval between the two events. By the second of (14.12) we have

$$dx_4=\gamma\, dx'_4 ,$$

or

$$dx_4/dx'_4=\gamma=(1-v^2)^{-\frac{1}{2}} . \tag{14.15}$$

This is clearly greater than unity. What does this mean in words? We must remember that the two events occur in the history of a *single* S′-clock, whereas they occur in the histories of *two* different S-clocks. But the S-clocks are synchronised, and S accepts dx_4 as the measure of relative time. If the S′-clock ticks once between the two events, then by (14.15) the S-clock ticks more than once. Hence it is natural for S to say that the S′-clock is running slow in the indicated ratio. This is *clock-retardation.*

This can be pretty confusing, because we are trying to fit relativistic ideas into a quasi-Newtonian way of looking at things. In the backs of our minds there hover two dangerous thoughts. First, the thought of a rigid body – rigid bodies cannot change their lengths, so how can a moving body contract? The answer to this is that *there are no rigid bodies*, only clock-clouds, and the measurement of distance is not something to be done with a 'rigid' ruler, but an elaborate matter involving light-signals. The second dangerous thought is that there is an absolute time, recorded by any perfect clock. Here the answer is that *there is no absolute time*, and that the clocks in the clock-clouds are doing their best, each recording its own personal time.

Since seeing is believing, it would be nice to display the contraction and retardation of a moving clock-cloud in the form of a space-time diagram, but this is not easy. Up to this point the space-time pictures were merely qualitative, more like cartoons than photographs, designed, like any rough sketch, to help us to concentrate our thoughts. There was no question of taking a ruler and making measurements on the picture.

But now we would like quantitative pictures, for we have precise values for the contraction and retardation and we would like to see these values displayed graphically. However, there is a certain difficulty in making quantitative space-time diagrams, as I shall now explain.

If we suppress the coordinates x_2 and x_3 as we have done above, an event is specified by two coordinates and we may represent that event on a sheet of paper by drawing rectangular axes and marking in a dot to represent the event. If A and B are two events in the history of a clock, we can mark in two dots to represent them. Suppose their coordinates are (x_1, x_4) and (x_1+dx_1, x_4+dx_4); in this work the d's need not be infinitesimal – they may be finite. Now the most important thing about those two events is the time-interval dt between them as recorded by the said clock, and for it we have the Minkowskian line-element

$$dt^2 = -dx_1^2 + dx_4^2 \,. \tag{14.16}$$

We would like this dt to be obvious when we look at the two dots on the paper.

But that is not what we see. Used as we are to Euclidean geometry, what we see is the distance AB, as measured by a ruler on the paper, and given by the Euclidean line-element

$$AB^2 = ds^2 = dx_1^2 + dx_4^2\,, \tag{14.17}$$

by the theorem of Pythagoras. On account of the minus sign in (14.16), this is quite different from the Minkowskian line-element (unless dx_1 happens to vanish), and we would be making a gross mistake if we were to identify the distance AB (measured by a ruler) with the time-interval (say t_{AB}) measured by a clock which includes the events A and B in its history. If we were reared on Minkowskian geometry we might be able to interpret the space-time diagram intuitively, but it is Euclidean geometry that is built into our systems and we find it much easier to think in Euclidean terms.

So I would say that if you want to grapple with the contraction and retardation of a moving clock-cloud, the best plan is to use the algebra as given above. It says everything there is to say. But as it says it in cryptic algebraic form, there is no harm in playing with space-time diagrams, which may serve to make things clearer, provided they are treated with due caution.

In dealing with the clock-clouds of the observers S and S′, it is wise to treat them fairly and not give undue privilege to one of them. Accordingly I propose to use *two* space-time diagrams, one for each observer. Let us then draw two pairs of rectangular axes as in fig. 14.1, with axes Ox_1x_4 for S and axes $O'x'_1x'_4$ for S′. Then any event appears *twice*, once in the S-plane and once in the S′-plane – with coordinates (x_1, x_4) in the first case and coordinates (x'_1, x'_4) in the second. The two pairs of coordinates are connected, as we have seen by the Lorentz transformation:

$$x'_1=\gamma(x_1-vx_4), \qquad x'_4=\gamma(-vx_1+x_4), \tag{14.18}$$

or equivalently

$$x_1=\gamma(x'_1+vx'_4), \qquad x_4=\gamma(vx'_1+x'_4), \tag{14.19}$$

where v is the (constant) velocity of S′ relative to S and

$$\gamma=(1-v^2)^{-\frac{1}{2}}. \tag{14.20}$$

The formulae (14.18) map the S-plane on the S′-plane. That is, given any point in the S-plane, the formulae give a corresponding point in the S′-plane. Likewise the formulae (14.19) map the S′-plane on the S-plane. The mapping is very simple. Since our aim is to be quantitative, it is a good idea to choose a value for the velocity v and draw diagrams to scale. Now we can give v any value between -1 and 1, and we want a value such that the diagram does not run off the page (as it would if we made v close to -1 or to 1) and such that it is not dull (as it would be if we made v very small – real life is dull, relativistically speaking, because most relative velocities are very small compared with the velocity of light). I decided to take $v=\frac{1}{2}$, so that the relative velocity is half that of light, and and the result lies comfortably between the two undesirable extremes. With $v=\frac{1}{2}$, (14.20) gives $\gamma=1.15$, $\gamma^{-1}=0.87$.

When we apply the transformation (14.18) to the rectangular grid in the S-plane shown in fig. 14.1, we get the rhombic grid in the S′-plane shown in the lower half

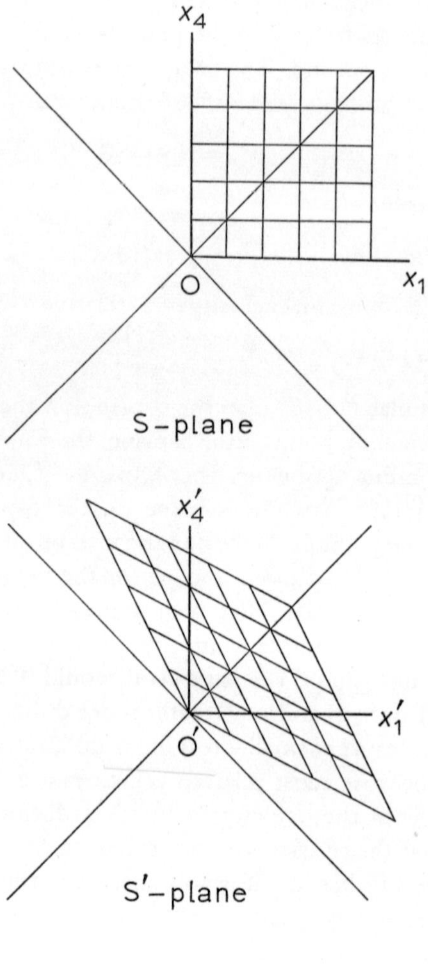

Fig. 14.1
Square S-grid maps into rhombic S′-grid

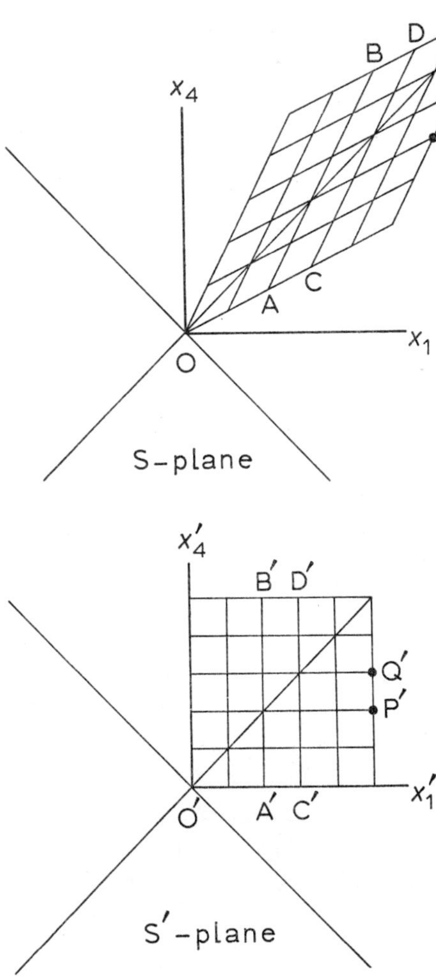

Fig. 14.2

Square S′-grid maps into rhombic S-grid

of fig. 14.1. If we want to find where any event goes to, it is merely necessary to identify the particular mesh of the grid. You might think that it would be possible to effect this transformation by playing with a lattice of rods pinned together, but this would not do, because although each square is deformed into a parallelogram, the (Euclidean) lengths of the sides are increased.

The lines inclined at 45^0 to the axes represent the histories of light-signals; they are unchanged by the mapping.

In fig. 14.2 the transformation (14.19) is applied to map the rectangular S′-grid into a rhombic S-grid.

Let us see what light these diagrams throw on the contraction and retardation of a moving clock-cloud.

For contraction, we are to follow two S′-clocks. Their histories appear in the lower part of fig. 14.2 as two lines parallel to the $O'x'_4$ axis – say the lines A′B′ and C′D′. The events on these lines appear in the upper part as the oblique lines AB and CD. Now the game S plays is to measure the distance between these lines for some given S-time, i.e. for some given value of x_4. So what you have to do is this: take a ruler, lay it parallel to Ox_1 and measure the distance between the lines AB and CD. This distance is not the same as the distance B′D′, but smaller – that is the contraction. How much smaller? Smaller in the ratio $1:\gamma$, that is 0.87. Try it yourself. If the diagram is too small to allow this order of accuracy, take a large sheet of squared paper and work to a larger scale. If you don't come up with the ratio 0.87, there's something wrong somewhere.

As for retardation, we are to take two events in the history of an S′-clock, say P′ and Q′ in the lower part of fig. 14.2. Those same two events appear as P and Q in the upper part (the S-plane). The observer S is to measure the time-interval between them, using his synchronised clocks (he must use two clocks because the two events P and Q do not occur in the history of any one of his clocks). This means that you are to take a ruler and measure, not the length PQ, but the projection of that length on the Ox_4-axis. In other words, draw a right-angled triangle with sides parallel to the axes and with PQ for hypotenuse; measure the side parallel to Ox_4. You should find it to be greater than P′Q′ in the ratio $\gamma = 1.15$. If you don't believe me, try it for yourself on a larger scale. This is the retardation.

We have been working with fig. 14.2 without using fig. 14.1, but that is because we have been thinking of observations made by S on the clock-cloud of S′. If you want to do it the other way round, then you should use fig. 14.1, and you should find the same results:

$$\text{contraction} = \gamma^{-1} = 0.87\ ,$$

$$\text{retardation} = \gamma = 1.15\ .$$

Incidentally, the rhombic lattices in the two figures are the same, except that one is turned through a right angle. I need hardly say that I have drawn only representative parts of the several lattices – they cover the whole plane.

The contraction (often referred to as the FitzGerald–Lorentz contraction) is not of much interest physically. The phenomena which it was originally invented to explain are better explained in other terms. But the retardation is of great physical interest. Some elementary particles have a statistical half-life, measured in terms of a 'clock' carried by the particle. When viewed by an experimenter, relative to whom the particle has a velocity v, the half-life will appear to be increased in the ratio γ:1. Thus if the true half-life is 1 second and the velocity half that of light ($v=\frac{1}{2}$), the half-life will appear to be 1.15 seconds, and if the velocity is nearly that of light, the observed half-life is very great.

A photon is in a rather peculiar position with regard to half-life. Since a photon clock does not tick at all (p. 111), it would seem that its true half-life (if it had one) must be zero. On the other hand, for a photon we have $v=1$ and γ infinite, so that the observed half-life should be infinite. This last conclusion seems to be correct, physically. A photon appears to be a stable particle.

CHAPTER 15

Particles in collision

Fig. 15.1 is a qualitative picture of flat space-time. The line-element is

$$dt^2 = -dx_1^2 - dx_2^2 - dx_3^2 + dx_4^2 . \tag{15.1}$$

The curve C is the world line of a particle and A is an event on C. The coordinates of A are x_i and those of an adjacent event B on C are $x_i + dx_i$, so that dx_i are the components of a small *4-vector* tangent to C. (We call it a 4-vector to remind us that it has *four* components which transform according to a Lorentz transformation if we change the frame of reference.)

Let dt be the time-interval between A and B as recorded by a clock carried by the particle. It is connected by (15.1) with the small 4-vector dx_i. On dividing dx_i by dt, we get a finite 4-vector dx_i/dt, and this we call the *4-velocity* of the particle at the event A. It may be denoted by V_i or by the single letter $\boldsymbol{V}$. Note that $\boldsymbol{V}$ is tangent to C at A. Do not confuse this 4-velocity $\boldsymbol{V}$ with the velocity v of equation (14.10).

As for the Minkowskian magnitude of $\boldsymbol{V}$, division of

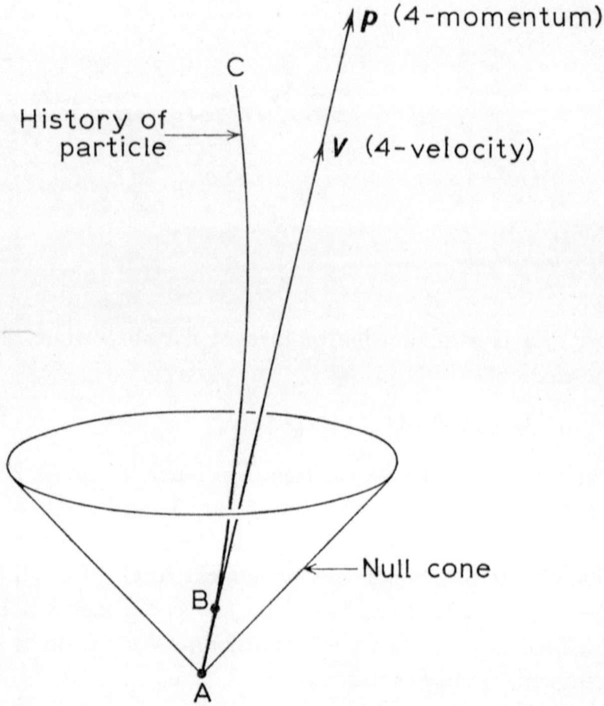

Fig. 15.1

4-velocity and 4-momentum of a particle

(15.1) by dt^2 gives

$$\boldsymbol{V}^2 = -V_1^2 - V_2^2 - V_3^2 + V_4^2 = 1 .$$

Thus the 4-velocity of a particle is a *unit vector*. Since the four components are connected by the last equation, only three of them can be freely chosen.

The vector $\boldsymbol{V}$ points into the future and must lie inside the null cone, as indicated in fig. 15.1 (no particle can travel as fast as light!).

The values of the four components V_i depend on the frame of reference used, and when we change the frame they transform by a Lorentz transformation. For any frame there are four axes of the coordinates x_i, and a Lorentz transformation may be regarded as a rotation of those axes, this 'rotation' being in some ways like ordinary rotations in Euclidean 3-space, but with the sort of distortion exhibited in the preceding chapter. By such a rotation we can turn the x_4-axis in the direction of the 4-velocity $\boldsymbol{V}$, and, if this is done, the first three components vanish and we have

$$V_1 = V_2 = V_3 = 0, \qquad V_4 = 1 . \tag{15.2}$$

By this trick we 'reduce the particle to rest' at the event A. But of course we must be careful never to say 'This particle is at rest' without specifying what frame of reference is being used.

So much for 4-velocity. It is a concept that belongs to *kinematics*. To deal with *dynamics*, we need a new concept, the *mass* of a particle. This concept is not much different from the Newtonian concept of mass. It is a positive number belonging to any particular particle and remains constant during the history of the particle, unless the particle experiences a catastrophe such as a collision with another particle. In a collision particles may change into new particles with new masses, with perhaps the production of photons with zero mass (see below). I shall denote the mass* of a particle by m.

If we multiply the 4-velocity $\boldsymbol{V}$ by the mass m, we get a new 4-vector pointing in the same direction in space-time. This is called the *4-momentum* of the particle and is denoted by p_i or $\boldsymbol{p}$. Thus we have

$$\boldsymbol{p} = m\boldsymbol{V} \,. \tag{15.3}$$

In the physics of elementary particles this 4-momentum is more important than 4-velocity, because there is a law of conservation of 4-momentum, of which more later.

But now a word about photons. A photon has a world line, but on it $\mathrm{d}t = 0$ ('photon-clocks do not tick'). Hence if we try to construct the 4-velocity of a photon as we did for a particle by writing $V_i = \mathrm{d}x_i/\mathrm{d}t$ we run into a bit of nonsense because we cannot divide by zero. Thus a photon has no 4-velocity. But it has a 4-momentum $\boldsymbol{p}$, which may be regarded as a limit of (15.3) when m tends to zero and the components of $\boldsymbol{V}$ to infinity so that a finite $\boldsymbol{p}$ results. Thus the mass of a photon is zero**. The same limiting process suggests that the 4-momentum $\boldsymbol{p}$ of a photon must point along a generator of the null cone, corresponding to motion with the speed of light.

We can now forget about the world line C of fig. 15.1

* This is often called *proper* mass or *rest* mass to distinguish it from *relative* mass, but I shall not be talking about relative mass, and so there should be no confusion.

** This applies also to the elementary particle called the neutrino, but I shall talk only of photons.

and concentrate on 4-momenta, which we picture as vectors in a four-dimensional space (*4-momentum space*) as in fig. 15.2 which shows the 4-momentum $\boldsymbol{p}$ of a typical particle and the 4-momentum $\boldsymbol{p}$ of a typical photon. The null cone shown here consists of the 4-momenta of all possible photons.

Since the 4-velocity $\boldsymbol{V}$ of a particle is a unit vector, it follows that the magnitude of its 4-momentum $\boldsymbol{p}$ is equal to the mass m. On the other hand, since the vector $\boldsymbol{p}$ of a photon lies on the null cone, its magnitude is zero. (Here and later, these magnitudes are Minkowskian magnitudes, not the ruler-lengths of the diagram.) Thus we have

$$\begin{aligned} &\text{for a particle: } -p_1^2-p_2^2-p_3^2+p_4^2=m^2\,, \\ &\text{for a photon: } -p_1^2-p_2^2-p_3^2+p_4^2=0\,. \end{aligned} \tag{15.4}$$

In Newtonian mechanics, the kinetic energy of a body is $\frac{1}{2}mv^2$, where m is the mass and v the speed, and this kinetic energy is a very real thing. If a car collides with a lamp-post, a crude but effective way of estimating the probable damage is to calculate the kinetic energy of the car. The relativistic energy which we shall now define is just as real in the microscopic world of elementary particles and intrudes into the macroscopic world in the form of atomic energy.

The *energy* E of a particle or photon is the fourth component p_4 of its 4-momentum:

$$E=p_4\,. \tag{15.5}$$

In some ways this is a queer definition. Why pick out one of the four components and give it a special name? Further, the value of p_4 depends on the frame of reference. However, if we rewrite the first of (15.4) as

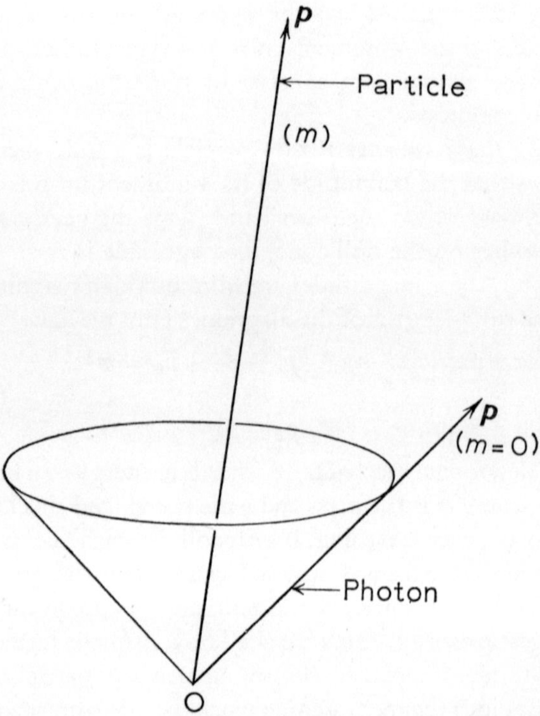

Fig. 15.2

4-momentum space

$$E^2 = m^2 + p_1{}^2 + p_2{}^2 + p_3{}^2\,, \qquad (15.6)$$

an interesting fact emerges. For each of the four terms on the right-hand side is *positive* and so the energy E always exceeds the mass m, except when we use the rest-frame of the particle; in that case $p_1 = p_2 = p_3 = 0$ and so

$$E = m\,. \qquad (15.7)$$

This is one of the most famous formulae in physics. It is Einstein's formula $E = mc^2$ where c is the speed of light, here simplified because I have not been using separate units for space and time; I have based the measurement of distance on that of time, so that the 'speed of light is unity' or $c = 1$.

It would be quite wrong to apply (15.7) to a photon by putting $m = 0$ and deduce that the photon has zero energy. For to get (15.7) we had to rotate the space-time axes (by a Lorentz transformation) so that the fourth axis had the direction of the particle's 4-momentum. That cannot be done for a photon, for we cannot make an axis lie on the null cone. So there is nothing to do in the case of a photon but write

$$E = p_4 = (p_1{}^2 + p_2{}^2 + p_3{}^2)^{\frac{1}{2}}\,, \qquad (15.8)$$

and leave it at that*.

Problems about the collision of particles stand in the forefront of present-day research in theoretical physics, but it is a matter which I am not competent to discuss in detail, and besides such detail would be out of place

* Another famous formula might be mentioned here: $E = h\nu$ where h is Planck's constant and ν the frequency of the photon when considered as a wave. This is consistent with (15.8) because, under Lorentz transformation, frequency transforms like the fourth component of a 4-vector.

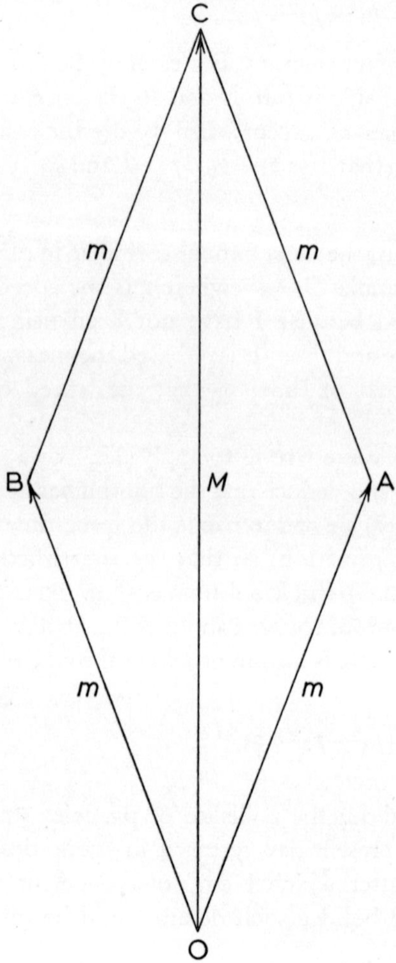

Fig. 15.3

A simple collision with coalescence

here. All I wish to do is to present some essentials in simple form for the better understanding of the relativistic principles involved. It is a curious fact that the source of atomic energy lies in a topsy-turvy triangle inequality. In Euclid we learn that a side of a triangle is *less* than the sum of the other two. For some triangles in space-time (the ones we are particularly interested in) the inequality works the other way round; one side is *greater* than the sum of the other two.

The basic law governing collisions is the law of *conservation of 4-momentum*, no matter how many particles or photons are involved and no matter whether there is transmutation from one kind of particle to another. This law states that

The total 4-momentum is unchanged by the collision.

By 'total 4-momentum' we mean the sum of 4-momentum vectors, the addition being carried out by adding components, or equivalently by the parallelogram law as in the elementary parallelogram of forces.

Consider now this simple case. Two particles, each of mass m, collide and coalesce to form a single particle of mass M. What happens?

For ease of reference, I shall call the incident particles A and B, and the final particle C. In fig. 15.3 the vectors OA and OB represent the 4-momenta of A and B, each with Minkowskian magnitude m. To find the total 4-momentum we have to add these two vectors and this we do by completing the parallelogram. Then the diagonal OC represents the 4-momentum of C. We have

$$\mathrm{OA}=m, \qquad \mathrm{OB}=m, \qquad \mathrm{OC}=M. \tag{15.9}$$

If we think in terms of components, and denote the 4-

momenta of A, B and C by $\boldsymbol{p}$, $\boldsymbol{p}'$ and $\boldsymbol{P}$ respectively, the law of conservation gives the four equations

$$P_1 = p_1 + p'_1, \qquad P_2 = p_2 + p'_2,$$

$$P_3 = p_3 + p'_3, \qquad P_4 = p_4 + p'_4 . \tag{15.10}$$

So far the frame of reference has not been specified. Let us use the rest frame of C. Then we have

$$P_1 = P_2 = P_3 = 0, \qquad P_4 = M , \tag{15.11}$$

and the last of (15.10) gives

$$M = E + E' , \tag{15.12}$$

where E and E' are the energies of A and B. But by (15.6) we have

$$E > m, \qquad E' > m,$$

and so

$$M > m + m' . \tag{15.13}$$

The mass of C is greater than the sum of the masses of A and B.*

Here is the topsy-turvy triangle inequality! In Minkowskian measure, the side OC in fig. 15.3 is greater than the sum of the sides OA and AC.

What about it? It is a fact of life in relativity and we have to accept it. But it does restrict what A and B can do by hurling themselves together. By intensifying their attack, they can manufacture (at least in principle) a C with any mass up to infinity, but no matter how mildly they ap-

* This fact may be discussed in a crude way in Newtonian terms. A and B are heated by the collision, and this heat accounts for the excessive mass of C.

proach each other, they can never make a C with mass less than the sum of their masses.

The essential process involved in the hydrogen bomb and in the generation of energy in the stars is the conversion of hydrogen into helium, four hydrogen particles forming one helium particle. Although this is an over-simplification, we may think of this as happening in one collision, and, as the above argument may be used for four colliding particles as well as for two, we might conclude that the mass of the helium particle must be greater than four times the mass of a hydrogen particle. But it is not so. The mass of the helium particle is known to be *less* than four times the mass of the hydrogen particle.

This needs looking into, but instead of dealing with that complicated and rather artificial case, let us consider the collision of a proton and a neutron. They combine to form one deuteron. The situation is similar, since it is known that the mass of a deuteron is *less* than the sum of the masses of a proton and a neutron.

In fig. 15.4, OB and BC represent the 4-momenta of the proton and the neutron, and so OC represents total 4-momentum. By the topsy-turvy triangle inequality, OC is *greater* than the sum of the masses of the proton and the neutron and so, if we measure off on OC a Minkowksian distance equal to that sum, there is a bit over at the top. Now the deuteron mass is *less* than that sum, and so we can accommodate within OC a vector OD to represent the 4-momentum of the deuteron. In fact, the correct total 4-momentum can be supplied by a deuteron with 4-momentum OD plus something else with 4-momentum DC. This last represents energy released by the collision.

However fig. 15.4 is not realistic. Elementary particles

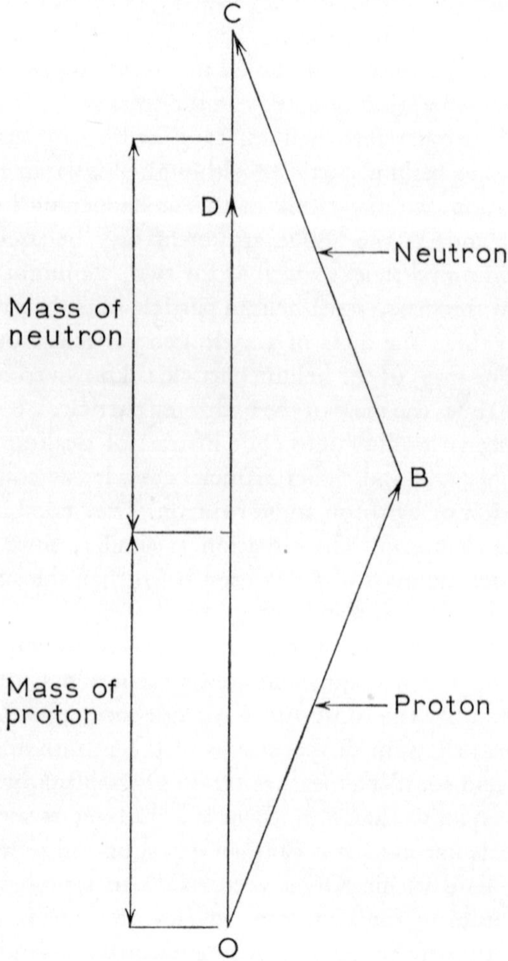

Fig. 15.4

A proton and a neutron turn into a deuteron plus something

have certain definite masses and it is unlikely that there would be a particle with mass equal to DC. We get a more plausible diagram by swinging the 4-momentum of the deuteron out of the line OC and filling the gap with a null vector DC representing the 4-momentum of a photon. Thus, as depicted in fig. 15.5, a proton and a neutron turn into a deuteron and a photon, and the energy of the photon may be used for useful or destructive purposes. This is perhaps the simplest example to illustrate the principles of the generation of nuclear power, and the source of this power is indeed the topsy-turvy character of the triangle inequality in Minkowskian space. Perhaps I should mention that a proton is the (charged) nucleus of a hydrogen atom, a neutron is that nucleus deprived of its charge, and a deuteron is built out of a proton and a neutron. The nucleus of the helium atom is built of two deuterons.

I want to bring this book to an end here because I have talked enough. If this were a real dialogue, you would chip in with some query or some constructive suggestion, I would reply to the best of my ability, and this might continue to the going down of the sun and later. But a book must end without dialogue of that sort, and I feel that this is where it should end. My aim has been to explain the concepts of relativity with the minimum of mathematics and at this stage I have reached a point somewhere between success and failure. The next step, if any, is up to you. If you think you understand the essential concepts sufficiently well (one never understands concepts perfectly), then you are in a position to grapple with the mathematics of relativity, some of it fairly easy and some of it of great difficulty. On the other hand, if you feel as puzzled as you

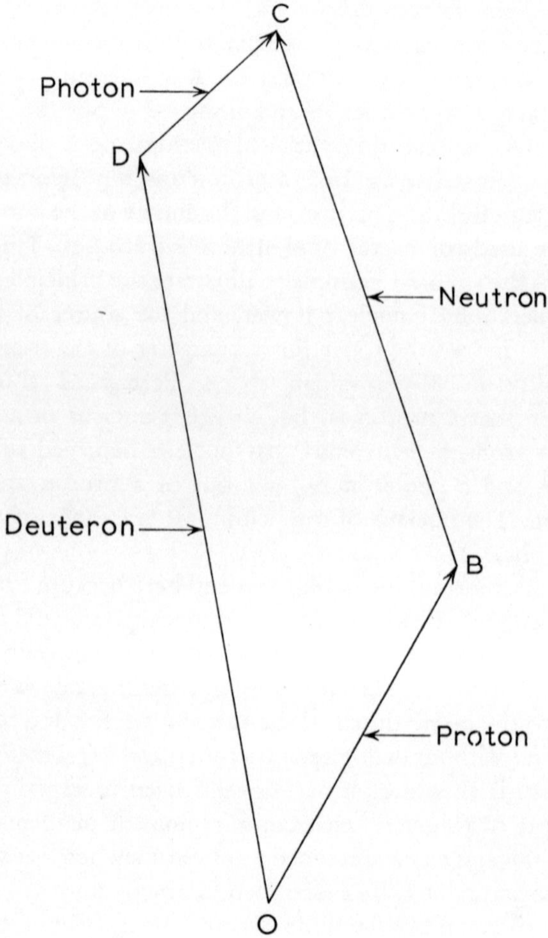

Fig. 15.5
A proton and a neutron turn into a deuteron and a photon

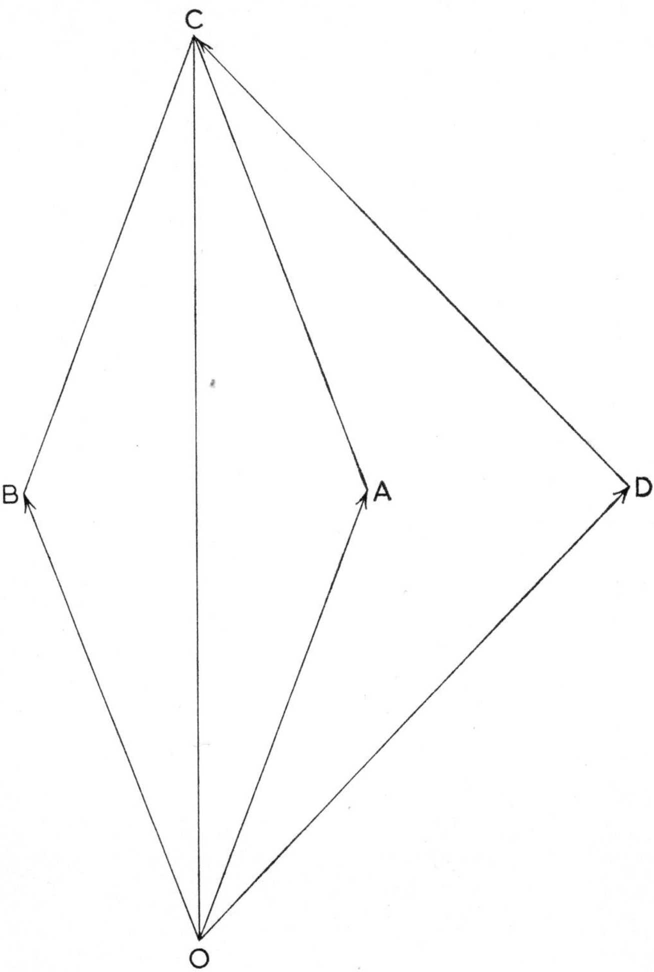

Fig. 15.6

Annihilation or creation of matter

ever were, you can tackle some of the many books and papers devoted to relativity and perhaps develop your own concepts along other lines.

As a farewell exhibit, I would like to show how matter is annihilated or created, subject to the law of conservation of 4-momentum. In fig. 15.6, OA and OB are the 4-momenta of two particles which collide. The total 4-momentum is the vector OC, obtained by the parallelogram rule. The two particles cannot simply disappear, leaving no trace, because, to balance the books, whatever emerges must give that same total 4-momentum, OC. Tired of their material existence, the particles would like to turn into a photon, but the law of conservation prohibits this – the 4-momentum of a photon is a null vector (of zero Minkowskian magnitude) while OC is not a null vector. But they can turn into *two* photons, such as those with 4-momenta indicated as OD and DC. That is what happens. That is how an electron and a positron are annihilated, changing from matter into two quanta of electromagnetic radiation. Or it might be the other way round. Two photons with 4-momenta OD and DC may turn into two particles, and so matter is created out of electromagnetic radiation. The details of such annihilation and creation lie far outside my scope; all I am trying to do is show how all such catastrophes are subject to the law of conservation of 4-momentum and how the consequences of that law may be exhibited in a simple diagram.

Note, incidentally, that fig. 15.6 illustrates the topsy-turvy triangle inequality: OC is positive and OD and DC are zero.

Index